MW00850035

The Nature and Nurture of
Narcissism

Understanding Narcissistic Personality Disorder from the Perspective of Gene-Environment Interaction

Peter Salerno, PsyD

Copyright © 2024 by Peter Salerno

First Edition

The Library of Congress has cataloged this paperback edition. Publication data available upon request.

ISBN: 979-8-218-40057-6 (paperback)

Printed in the United States

Thank you for purchasing an authorized edition of this book. Any part of this book may be used in any form, using normal citation practices. All Rights Reserved.

DSM-III and *DSM-5-TR* are registered trademarks of the American Psychiatric Association. *DSM* general personality disorder criteria and criteria for narcissistic personality disorder were reprinted with permission.

The information in this book is intended for general inquiry and informational purposes only and should not be utilized as a substitute for medical advice, diagnosis, or treatment. If you think you or those under your care are ill or in need of health care of any kind, please seek immediate medical attention. Always consult with a doctor or other competent licensed clinical professional for specific recommendations about medical treatments for yourself or those under your care. Any use of, or reliance in any way upon, the information contained in this book is solely at your own risk.

The author has checked with sources believed to be reliable in his efforts to provide information that is complete and generally in accord with the standards of practice that are accepted at the time of publication. However, in view of the possibility of human error or changes in behavioral, mental health, or medical sciences, neither the author nor any other party who has been involved in the preparation or publication of this work warrants that the information contained herein is in every respect accurate or complete, and they are not responsible for any errors or omissions, or the results obtained from the use of such information. Readers are encouraged to confirm the information contained in this book with other sources.

CONTENTS

About the Author

Peter Salerno, PsyD, LMFT, is a licensed psychotherapist and award-winning author. He holds a Doctor of Psychology degree (PsyD), a Master of Science degree (MS) in Clinical Psychology, and a Bachelor of Arts (BA) degree in English Literature. Dr. Salerno is a trauma specialist, clinical supervisor, and consultant who utilizes empirically validated, science-based approaches to promote healing and self-empowerment. Dr. Salerno is trained in Eye Movement Desensitization and Reprocessing Therapy (EMDR) and is certified in Family Trauma and Complex Trauma through the International Association of Trauma Professionals. He also holds certifications in Psychoanalytic Psychotherapy and Personality Disorder Treatment (C-PD) and is trained and qualified to administer and score the Hare Psychopathy Checklist-Revised (PCL-R). Dr. Salerno has treated mental health conditions in a variety of clinical settings and has authored six books on a broad variety of topics, including stress and trauma, early development and attachment, mind-body integration, emotion and cognition, personality pathology, and philosophy. Dr. Salerno currently works with individuals, couples, and families in private practice. He resides in Southern California.

Website: drpetersalerno.com
Email: peter@drpetersalerno.com
Instagram: @drpetersalerno

Dedication

For my mother and my grandmother.
I am grateful for you both, for completely different reasons.

Preface

*A devil, a born devil, on whose nature
Nurture can never stick . . .*

—William Shakespeare,
The Tempest

When it comes to the topic of narcissism, there are a lot of people, me included, who enjoy literature that goes for the jugular. This book is one of them, but not right away. First, we will need context. A brief history of narcissistic personality disorder in the field of mental health is required before we can fully engage in the drama that is narcissism and learn, ultimately, what we can do about its threat in our lives. To understand where we are at in terms of the most current research on narcissism, I want to invite you on a trip down memory lane—specifically *my* memory lane—so please bear with me; I promise the trip will be worth taking.

Chapter One
All in the Family?

A Grandmother's Tale

When I was a child, I used to wish that my mother could bring herself to hate her own mother. It was an odd wish for a young child to have during the first decade of life, especially because I loved my maternal grandmother and she treated me kindly. It wasn't until decades later when I began studying clinical psychology that my perplexing childhood wish began to make sense. Nobody in my family had a clue that my maternal grandmother—may her impossibly complex and menacing soul rest in peace—met the full criteria for two severe personality disorders: narcissistic personality disorder and borderline personality disorder.

Talk about hell. No one in our family knew much about mental health, let alone mental disorders. In the 1980s and '90s, people who had "mental problems" were commonly assumed to be suffering from psychotic conditions like schizophrenia, nervous breakdowns (which aren't really a thing), PTSD, depression, anxiety, or manic depression, now known as bipolar disorder, all of which are as highly misdiagnosed and misunderstood today as they were decades ago.[1]

But that was pretty much par for the course in terms of mainstream awareness of mental illness. The diagnoses and features of personality disorders were not common knowledge, and no one in my family had a background in psychiatry or psychology, so no one had any reason to suspect my grandmother was anything other than someone who vacillated between acting crazy and acting normal or between acting cruel and acting kind. If that wasn't enough, my grandmother constantly required attention and admiration—excessively so. In addition, she was vain; felt entitled; was deceptive, manipulative, and lacking in empathy; and believed she deserved special treatment regardless of the circumstances. She was always right, and you were always wrong, end of discussion. My family members tended to chalk this up to culture: after all, everyone knows that all Italians are loud and dramatic and have poor boundaries, right? Wrong.

---

[1] M. Bloom, *Mastering Differential Diagnosis with the DSM-5-TR: A System-Based Approach* (Pesi Publishing, 2024).

As a child, my maternal grandmother's personality "traits" were baffling to experience firsthand. As an adult studying and exploring human behavior, the dichotomy I had personally experienced proved fascinating. On the one hand, I was treated like a saint by my grandmother. She doted on me and showered me with money, toys, candy, and compliments. She sang to me, rocked me on her lap, and made me feel safe and happy . . . up until around the age of seven. I know now that I was treated so well by my grandmother because as a child with a highly agreeable temperament, I mirrored her perfectly. Had I disagreed with her or challenged her, I'm certain that things would have been different for me.

By contrast, during the first seven years of my life, my grandmother treated my mother in a completely opposite manner, which I bore witness to on a constant basis. My grandmother devalued my mother, demeaned her, ridiculed her, shamed her, guilt-tripped her, hurled insults at her, screamed bloody murder at her, and fiercely berated her in front of me at the drop of a hat. One minute my close-knit nuclear family and I were enjoying a home-cooked meal at Grandma and Grandpa's house; the next minute, my Italian grandmother, unprovoked, would erupt like Mount Vesuvius, launching a tirade of harsh insults at my mother in both English and Italian—names that, to respect the sensitivity of many readers, won't be repeated here. But you can use your imagination. If a word could hurt, my grandmother would use it.

And worst of all, this explosive barrage was not something that happened on rare occasions— it was more a common occurrence than not. And too often, dumbfounded, I bore witness to it. How did my mother behave in response? With quiet dignity. She would sit and take these verbal lashings with stoicism, and then, when the storm abated, my mother would quietly gather our belongings and we'd all leave.

Then it would happen all over again, nearly every time we'd visit my grandparents. And we visited a lot. My grandmother would verbally and behaviorally idealize me while publicly devaluing my mother. The few times my mother stood up for herself and pleaded with my grandmother not to behave like that in front of me—her "favorite" grandson—the crocodile tears would begin to flow down Grandmother's cheeks. She self-morphed into the injured party, a victim of what she deemed my mother's unreasonableness, reacting as if someone had literally just stabbed her both in the back and in the heart. Accusations of betrayal would ensue. Sobbing and self-victimization would follow: "Nobody cares about me. Everyone has abandoned me. I'm all alone. I have no one. I'm in such despair."

All alone? No one? Abandoned? Even to a child that seemed like an odd thing to say, considering everyone in our family bent over backward for my grandmother's sake at all times, no matter what was going on or how inconvenient it was. She was treated like royalty. Her family members all catered to her like servants in her imperial court, yet she claimed to be mistreated and neglected ALL THE TIME. Nothing my mother—or later, the rest of my family—ever did was ever good enough, or right enough, or even just okay, much less appropriate or helpful. Everyone, it seemed, according to grandmother, was a terrible sinner who fell short and failed her.

The aftermath, too, following one of grandmother's outbursts, followed a pattern. My tenderhearted mother would comfort my grandmother, even though she'd just been the target of a torrent of unspeakable words a mother should never call her daughter nor should ever be uttered in the presence of young grandchildren. Then, as if by some divine intervention, my grandmother would suddenly become totally agreeable. Not only that, but she would also "forget" what she had just said and done and would insist that it never happened. How convenient. For *her.* For the rest of us who had suffered through her outburst, not so much.

Like many people who meet the criteria for borderline personality disorder, the only thing consistent about my grandmother's thoughts, feelings, and behavior was inconsistency. She was impulsive, excessive, reckless, self-destructive, and abusive to others in every sense of the word, but her power over others was that *she* held "the other shoe" that everyone in her orbit feared would drop at any given moment.

And, like many people who meet the criteria for narcissistic personality disorder, my grandmother was lacking in empathy, discarded the needs and feelings of others effortlessly, especially those of her daughter, my mother, displayed a grandiose sense of self-importance, and demanded, without question, confirmation of whatever she needed confirmed no matter how much it departed from reality.

My grandmother's vanity—mainly connected to her outward physical appearance—was off the charts. She was also an impulsive spender of money she didn't earn herself, and also a bit of a thief. After she died, we found over $900,000 in cash, all skimmed from my grandfather's income since the 1960s, that she had hidden and stashed in various cubbyholes throughout their house. My grandfather, in addition to the challenge of being married to my grandmother, was mired in his own personal struggles with alcoholism and military-related PTSD, and didn't even know the money had been siphoned from the family funds all those decades.

When my grandmother reached the age of forty, something inexplicable to everyone in our family happened: she stopped leaving the house. For good. For the last forty-three years of her life, my grandmother existed in self-imposed exile. Why? Not because she had developed a fear of leaving the house, but because of her vanity. In her mind, at forty, she had concluded she was not going to be young and flawless forever, so she was determined not to permit anyone to bear witness to her aging process. She couldn't bear to be seen in public as she grew older. Not ever again. It was *forbidden*. Nature had made it impossible for her to perfectly manage her own ideal image, so she retreated into the shadows. In her mind, wrinkles were unacceptable. Growing old was not natural; it was sinful.

Throughout my grandmother's life, no one was ever certain of her actual age. She regularly claimed to be younger than she was and would edit the year she was born depending upon who she was talking to. The mystique of eternal youthful beauty was her constant obsession, and she was willing to sacrifice years of her life in an effort to maintain the illusion she believed she'd created.

Years later, when I was in graduate school, I discovered that my grandmother had been heavily abusing prescription narcotics and anxiolytics as well as over-the-counter medications since I was a kid. Substance abuse is very common among narcissistic and borderline personalities. My grandfather, who was a chiropractor, had medical connections that enabled the constant supply of these "legal" drugs, which was fairly common a few decades ago. He knew doctors he could ask as a favor to prescribe my grandmother unlimited amounts of Ativan, Valium, Vicodin, and other drugs. Due to his enabling—whether well-intentioned or not—my grandmother was able to maintain her drug abuse, and borderline personalities and narcissists who abuse drugs are a lethal combination.

Why didn't my grandfather simply leave? Being attached to a borderline/narcissistic personality is similar to being caught in a mouse trap. When you're caught, it's excruciatingly painful and you are well aware that the entrapment could eventually kill you—literally. But at the same time, not only can you not move because you're trapped, but also most of the time, you *wouldn't* move even if you *could*. Inertia becomes a way of surviving.

The enmeshment and inertia are torture for someone who desperately wants a relationship with someone in the full throes of narcissistic and borderline personality disorder. Such a situation is excruciating for someone who badly wants to help a loved one to feel better and live a normal life. Someone who has been conditioned throughout their life to caretake a borderline/narcissist. Someone like my mom.

My mother spent a great deal of her childhood and adult life trying to atone for sins she never committed against my all-knowing, all-powerful, self-proclaimed "selfless" grandmother. While my mother may not admit it, guilt and shame still plague her today—unfounded, undeserved guilt and shame that never belonged to her. To this day, my mother still loves her mother more than ever, ten years after my grandmother's passing, and sometimes when we talk about what my mother has been through, I secretly wish she could see her mother for who she really was and not the person she wishes she had been. If only to hold her somewhat accountable beyond the periphery of her rose-colored memory. If only to get a break from the guilt and shame—just a little bit of a break at least.

In grad school, we all diagnosed our entire families back generations. We were told not to, but no one listens—the temptation is too great. We all pulled out the *Diagnostic and Statistical Manual of Mental Disorders* (the *DSM* for short) to try to make sense of our own baggage. As I became further immersed in the study of psychopathology, I became curious about the details of my own family history. Like a lot of people, I didn't really know much about my own lineage. What I discovered was that most of the "bad" things were tightly held secrets within the family. When I did manage to pry out information from relatives, I was offered contradictory information from nearly everyone I spoke to, which just confused me even further. Who knew the truth? Who would tell it? It was during this informal investigation into my own heritage in order to help me make sense not only of the way *my* mind worked, but also of my family dynamics, that I discovered my mother's childhood was worse than I could have ever imagined.

All my mother ever wanted was a loving connection with her mother. Someone to go shopping with, to talk with about friends and crushes. She wanted a maternal parent to invest interest in her. All my mother ever really wanted was a mother—a loving parental relationship that every human being should earn as their birthright. What my mother got was something completely different.

As a child, my mother endured public shaming and body shaming and was constantly subjected to verbal insults. Her mother slapped her in the face repeatedly, hit her with spoons and dinner plates, threw knives at her from across the room, and dragged her around the house by her hair.

But not all abuse is *physical*—abuse also comes in the form of non-empathetic responses. When my mother was five years old, she was lured into a garage by a neighborhood teenage boy, sexually assaulted, and locked in a wooden trunk. When my grandparents became aware that my mother was missing, they went door-to-door looking for her. They eventually found her in the trunk in the neighbor's garage because my mother, a little child only five years old, was banging on the lid of the trunk from the inside in an effort to call for help. When they opened the trunk, they found my mother huddled inside wearing only her underwear. They lifted her out and walked her home.

Instead of being horrified and outraged at what had happened to her daughter, my grandmother blamed my *mother* for being naked in the trunk. My grandmother didn't have even an ounce of sympathy or compassion for her child. She made her tearful and traumatized five-year-old sit naked on her twin bed and think about what *she'd* done. My mother remembers she'd wet her underpants due to the intense, unbearable terror and stress on her young nervous system—in the nauseating darkness of that old trunk, she had been certain she was going to die. Instead of trying to soothe her young daughter, my grandmother hung the soiled undergarment above the mirror of the dresser in my mother's childhood bedroom, as evidence of what my *mother* had done.

My grandmother was a real monster to my mother, yet my mother loved her with all her heart, unconditionally. Every opportunity she had, my grandmother attempted to steal her own daughter's self-esteem and self-worth. Why would a mother do this to their own daughter? Because narcissists loathe anyone whom they perceive as a threat to their own divine right to excessive admiration—including their own children. My grandmother viewed what had befallen my mother as a negative reflection on *her*, not as an injury to a small child in her care.

The Exception and the Rule

I don't "hate" my grandmother, and I'm not sharing this story in order to disparage her life or disrespect her as a human being. I do hate her disorders, and I am revolted at the way she chose to treat my mother. I purposely write "chose" because I do not believe in biological determinism or reductionism, and I do not believe that adverse environments make us helpless victims who cannot change. In my therapy practice, I've seen too many exceptions to these false assumptions. No one, including my grandmother, is a passive and helpless recipient of their genes or their environment. No one is.

When I became a therapist, I could clearly see that my mother met all the criteria for posttraumatic stress disorder (PTSD) as a result of what she'd experienced in early childhood. When I was growing up, I experienced my mother when she was in distress as irritable, defensive, hypervigilant, fearful, avoidant, and sometimes overly frustrated—all signs of someone with posttraumatic stress. But unlike her own mother, my mother was *never* cruel. She was *never* abusive. She was *never* neglectful. She *never* called me names, never hit me, never threatened me, never devalued or demeaned me. She was not a chip off the old block. She was very different than her own mother.

My mother had been through hell with her mother, but she didn't put her own offspring through hell in turn. She always did the best she could for everyone in our family—to a fault, in fact. She needed better boundaries for herself, but this was something she couldn't manage. Her frustration and irritation were not harsh—they were trauma responses fueled by fear and shame, *not* by a disorder in her personality.

And no matter what we might believe or what we've been taught, there is an undeniable difference between someone who behaves in certain ways because they are traumatized and someone who behaves in certain ways because they are personality disordered. They are *not* the same thing, and they do not have the same or even similar causes even if the behaviors themselves seem superficially similar.

Hold on! How can that be? Aren't borderline personality disorder and narcissistic personality disorder transmitted to the recipient of abuse as a result of cumulative trauma, whether complex trauma, developmental trauma, attachment trauma, or intergenerational trauma? Shouldn't my mother, because of *her* mother, be a full-blown borderline personality and/or a pathological narcissist?

Nope.

Despite her horrific, nightmarish upbringing, my mother's intact, stable, kind, and gentle personality is representative of millions of exceptions to the discredited myth and common thinking error that narcissism is *caused* by trauma or parenting style. Even though this theory prevails in popular culture and media today, it does not hold up in reality. Not at all.

There are millions upon millions of kind, loving, generous, flexible, responsible, and stable people who have narcissistic and abusive parents who they are nothing like. There are countless individuals who were abused and neglected but who are not narcissistic themselves by any means. How do we account for this phenomenon? If narcissistic parents who abuse and neglect their children, or who overindulge and overvalue them do not consistently produce narcissistic children who grow up to be narcissistic adults, then what *does* produce a narcissist?

That is what this book is going to reveal.

Science writ large has a strict policy about exceptions to theories: if there is even *one* exception to a theory, then science demands that the theory be revisited and revised.[2] An exception is an indicator that more research needs to be done. Better explanations need to be formulated. It's the rational and responsible thing to do when explanations don't add up and when reasons don't produce results.

And that's what this book is for—to follow the exceptions to their inevitable end, no matter where they lead.

A New Way of Viewing Narcissism

This book is a revision of something we have collectively and historically been wrong about. Too many of us have been indoctrinated by the false notion that *all* psychopathology is the result of a person's bad childhood, that bad childhoods always in some way hinder or immobilize the normal course of psychological

---

[2] Patricia A. Resick, Candice M. Monson, and Kathleen M. Chard, *Cognitive Processing Therapy for PTSD: A Comprehensive Manual* (The Guilford Press, 2017).

development, and that bad childhoods cause *all* mental disorders.

Well, guess what? Just like autism, schizophrenia, bipolar disorder, and ADHD were all theorized to be caused by parenting and have proven not to be, narcissism is also an exception to the so-called rule, even though this is not yet fully accepted as fact. The bottom line is this: there is a distinct difference between behaviors and tendencies that result from cumulative trauma, complex trauma, developmental trauma, attachment trauma, and intergenerational trauma that eventually develop into post-traumatic stress disorder, anxiety, or depression in an otherwise normally functioning and healthy person and pathological narcissistic traits, which are chronic in dysfunction, abnormal, do not develop from trauma, and are always present no matter what happens *because they are present from the start*.[3] Enduring, consistent characteristics and episodic symptoms that come and go are not the same thing. Not even close. It's surprising how obvious this is, yet how easily and frequently this reality has escaped us. Behavioral features of many conditions may resemble each

---

[3] J. Paris, *Myths of Trauma: Why Adversity Does Not Necessarily Make Us Sick* (Oxford University Press, 2023).

other and may even overlap, but the nature and the etiology—or cause—of the presence of narcissism are distinctly different.

This distinction became more and more apparent to me as I immersed myself in clinical psychology, psychiatry, neuroscience, and behavioral and molecular genetics while sitting with forty people each and every week in a clinical setting. My therapeutic practice has afforded me an up-close and personal look into the subtle and obvious differences in the presentation and personality of pathological narcissists and childhood trauma victims and survivors. Don't get me wrong: I am *not* saying that some narcissists have not been abused or neglected or that some haven't endured hell from others. Many have. But the reality is that just as many have not.[4]

This Book and You

If you are someone who has had the unfortunate circumstance of being raised by someone like my maternal grandmother, I hope this book alleviates any shame or guilt you do not deserve

---

[4] Stuart C. Yudofsky, *Fatal Flaws: Navigating Destructive Relationships with People with Disorders of Personality and Character* (American Psychiatric Association Publishing, 2005).

to carry, as well as helps you to think and believe differently about who you are, about what your value is, and about how you truly deserve to be treated. If you are someone who is married to or related to someone who is arrogant, grandiose, unreasonable, uncooperative, lacking in empathy, antagonistic, exploitative, dramatic, insensitive, enraged, entitled, disagreeable, selfish, uncaring, and impulsive, and *you* are blamed for their attitude or behavior despite doing all you can to provide them with a warm and loving relationship and environment, I hope this book will help you to realize you have *no* part in their pathology. If you are in such a situation, please understand that regardless of the peace and stability you are offering them, if they don't learn how to treat you better, then you might need to make a decision that is in *your* best interest, and that decision may require emotional and even physical distance.

If you are a parent who is at your wit's end because one of your children, no matter how well you treat them or how sensibly and responsibly you parent them and discipline them, simply will not respond appropriately, and your child's thoughts, feelings, and behaviors seem abnormal rather than age-appropriate, and may even seem to be getting worse, my desire is that this book provides an explanation that validates your struggle and permits you to stop blaming yourself while also providing you with hope. Because there is still hope.

My personal wish is that this book lends a helping hand so you can better recognize and accept your own good nature in contrast to the narcissist casting a shadow over your life, regardless of what anyone, including so-called experts, have said about you and your falsely presumed contributing role in the behavior of the narcissist in your life.

Finally, if you are a trauma victim or survivor (*not* synonymous or interchangeable terms) and you are tired of being grouped in the same category as a pathological narcissist when in your heart you know you are nothing like a narcissist, my hope is that this book dispels the myths that have contributed to any erroneous beliefs that you are a bad person, because trauma does *not* cause narcissism.

By the end of this book, you will know the true cause of narcissism, and it might surprise you. It may even confuse or disturb you, but once you get past the shock, I believe you might be relieved to know the truth. So buckle up and prepare to be astonished. Because almost everything that has ever been written in popular literature about narcissism . . . is wrong.

Chapter Two
Narcissists—Here, There, and Everywhere

Below is a series of brief introductions to a range of everyday people, some of whom might just seem painfully familiar. Because narcissism can be both pervasive and subtle, good people often give others with "difficult" personalities the benefit of the doubt when it comes to their abusive behavior. Since they tend to claim to have had difficult childhoods or to have experienced trauma in their pasts, we all want to believe that they deserve our sympathy and understanding. Or do they?

Case Study 1: The Contradictory Husband

A thirty-two-year-old man and a thirty-three-year-old woman who have been married for seven years come to my office for marital counseling. Throughout the initial session, the wife appears frightened. Her body language is guarded. She avoids eye contact, and unbidden tears run down her cheeks. She's five months pregnant and diabetic.

During our session, her husband gets choked up and even begins to weep openly. He tells me that when he was at church last Sunday, God spoke to him and told him to do whatever he could to save his marriage. Even though he says he doesn't "believe" in therapy, "God's orders are God's orders."

In the very next breath, his demeanor completely shifts. Before my very eyes, Dr. Jekyll becomes Mr. Hyde. He becomes enraged and tells me that if his wife ever sleeps in past 5:00 a.m. again "so much as once," he is going to serve her with divorce papers and seek custody of their unborn child. He then tells me that his mother was an alcoholic who, because she was passed out in the morning from excessive drinking in the night hours, neglected to take him to school in the morning when he was just a kid. He is not, he tells me adamantly, going to stand for his wife putting him through the same abuse that he suffered in childhood. He intends to nip his wife's "lazy behaviors" in the bud. He says the *only* reason he and his wife are having marital problems is because of her sleeping in past 5:00 a.m. During this tirade, his wife remains quiet and appears fearfully submissive.

The following week, the couple returns. The husband is in a very cheerful mood, as happy as a clam. He boasts about how he and his wife have been "boning" frequently and all is well at home.

"Praise God," he adds. His wife doesn't share his cheerfulness and doesn't look very comfortable when he inappropriately references their intimate life for really no good reason.

The wife proceeds to tell me that since our last session, her husband has insisted that she sleep in until 6:00 a.m. and has begun "surprising" her with breakfast in bed. But there is a catch: for her to be afforded these "luxuries," they have to have sex before it is time for him to go to work, and during sex, they will watch porn.

I inquire deeper to see what exactly has changed in the intervening week between our sessions. It turns out that the husband's shift at work changed from a 7:00 a.m. start time to a 9:00 a.m. start time.

It is his contention that his demands and contingencies are a result of his abusive and neglectful childhood—that he *has* to be controlling because he "needs" to ensure appropriate behavior in his home. Is this really the case?

Case Study 2: The Devaluing Detective

A thirty-eight-year-old law enforcement officer comes to my office for a mandated "fit for duty" evaluation. He shakes my hand with a smile and tells me that my office is not as big as he expected it to be. He sits down and starts scanning the room, stopping when he comes to the wall where my diplomas are displayed.

"A PsyD, huh?" he scoffs. "I was thinking about earning one of those in my sleep. It's not like it's rocket science. I mean, we all know psychology is not a legitimate science, bud. You should've been a cop. That's a *real* career. You get to shoot people, as long as they shoot at you first."

He's aiming for humor, but he's also serious. Since I remain silent and don't take the bait, he adds: "I know I can be an asshole. My dad was an asshole too. I guess I feel like I have something to prove because of him."

Did he really demean me and insult my credentials because his father was hard on him when he was a kid?

Case Study 3: The Promiscuous Bartender

A twenty-four-year-old female bartender comes to my office. Prior to her arrival, she reported to me over the phone that her boyfriend had abruptly moved out of their shared dwelling for no reason whatsoever and she's heartbroken by this turn of events.

She proceeds to tell me that she wants to do EMDR, or *eye movement desensitization and reprocessing*, a type of trauma treatment that targets and reprocesses traumatic memories, because she was sexually assaulted when she was a child. She believes the root of all her trouble stems from this trauma, in particular, her "low self-esteem." She believes her pervasive low self-esteem is why she regularly cheated on her recent boyfriend, as well as on every boyfriend she had in the past. She says she needs to learn some coping skills for her "numb and empty feelings" and needs to find a boyfriend who won't complain about her being out all night, won't be so jealous that older men hit on her all the time, and won't be threatened by the fact that all her drinking buddies are men.

During our session, she tells me that her boyfriend abused her and threatened her. While I can't be certain this is true after only five minutes, as a mental health professional I take her accusation seriously. I ask her if she is concerned about her safety. "Not at all," she scoffs. She proceeds to tell me that her boyfriend is a coward and confesses that she was skimming money from his bank account during the time they were together so that when he comes crawling back to her, she'll have enough money stashed away to be the one to walk out on him.

She tells me she's in love with another man even though he's a loser, but the sex is good. She says she's been very promiscuous and angry ever since her father kicked her out of the house at fifteen. Then she tells me she doesn't think she's ever really been in love because she gets bored with love too easily—it's sex that makes dealing with men worth it, especially when she's drunk or high.

Does this young woman think and behave this way because of unresolved childhood sexual abuse and paternal abandonment?

Case Study 4: The Nine-Year-Old Manipulator

A nine-year-old boy is brought to see me by his parents. He presents as very charming and smiles a lot in session, but when he talks about his "sad" experiences, his affect is incongruent with his words. His story seems a bit rehearsed and superficial, but I take him seriously and listen patiently.

Prior to the session, his parents told me over the phone that they were desperate. Their son is constantly disobedient and his tantrums quickly escalate to rage. They say he has dropped every friend he's ever had if they happen to have a toy or possession that he doesn't have, or if they win (and he doesn't) while playing a board game or video game. Furthermore, this nine-year-old constantly complains to his parents that they neglect him and ignore him and that because of them, he never feels good enough. I explore a bit and soon discover that his accusations of their inadequacies *only* erupt after they say "no" to one of his outlandish and unreasonable requests.

This mother tearfully tells me that she truly doesn't understand why her little boy feels the way he does. It seems like the more she gives her son and the more nurturing she is with him, the more entitled and exploitative he becomes. The father reports that the boy is also verbally and sometimes physically abusive to his sisters, who don't behave in a like manner whatsoever and who, at this point, don't want anything to do with their brother at all. The nine-year-old's parents express terrible guilt as they share these things and say they feel like bad parents for even reporting this to me.

As the young boy sits on my couch, a vape pen falls out of his sweatpants pocket. He looks to see if I saw this, then erupts in rage. "It's not mine!" he yells. Then, after I don't respond as he expected me to—with shock or concern or dismay—he smirks and says: "If you tell my parents,

I'll tell them that I told you my sister touches my privates in the bathroom but you didn't say anything."

He's nine years old. How do we explain such a level of arrogance, entitlement, and effortless manipulation in a little boy? Is it really because his parents are too giving? Too nurturing? Is this age-appropriate "acting out" behavior that he will eventually outgrow?

What to Make of This Array of Perplexing Personalities?

Why do such diverse individuals think and behave the way they do? Is it really because of how they were treated in childhood? Because of how they were parented? Or could there be another reason? Another cause?

What caused my maternal grandmother to be the way she was? Her environment? Was she traumatized beyond repair? Did she experience early childhood wounds? Developmental ruptures? If her environment as a child was similar to the environment that she exposed my mother to, was she just suffering from a repetition compulsion, forced to unconsciously replicate what she had experienced through no fault of her own?

Or could she have been a passive recipient of intergenerational trauma, trauma experienced by previous generations that was translated into an adaptation and transmitted to my grandmother, causing her to mistreat my mother the way she did? Did she not realize what she was doing? Did she not know better? If that's the case, why didn't my mother or her siblings do the same things to their children? Why wasn't my mother a thinking, feeling, and behaving carbon copy of her mom?

Or is the question better framed as *how* could my grandmother and mother be so different in personality and behavior?

The answers lie in the pages below.

Chapter Three
The Truth about Narcissism

It's more common than not for someone who hears the term *narcissism* to relate the cause to childhood adversity, whether it be from abuse, neglect, maladaptive parenting styles of any kind, or any other form of early hardship. That is precisely what I was taught in my degree programs. Contrary to the devaluing views shared by the detective in the case study about my education, earning a Doctor of Psychology degree is not an easy task that can be accomplished in your sleep. It is a rigorous commitment that spans years and requires discipline, fortitude, and intelligence. It also requires a dissertation process that is overseen by a committee of competent doctors, and the resultant dissertation must be defended in front of that same committee of doctors before any degree is awarded.

The Misinformed Therapist and the Allure of Pseudoscience

As a doctoral candidate or graduate student, quite simply, you are taught what you are taught by those who teach you. And I found out the hard way that what even the most reputable graduate psychology programs teach their students about the etiology—or cause—of narcissism is wrong. Not only that, but the root causes of the other nine personality disorders officially recognized and classified by the American Psychiatric Association are likewise incorrect. As much as it doesn't thrill me to point this out, graduate psychology programs across the country are teaching therapists-in-training about narcissism and other personality disorders, relying not on science-based facts but pseudoscience.

Many psychiatrists, psychologists, and psychotherapists, including me, can successfully earn a degree in our respective fields without obtaining a thorough understanding of the science that underlies the cause of narcissism and the other disorders we treat.[5] The truth is, the standards of training vary widely within a field like psychology, and empirical research and practice are not always valued as much as majority rule, faculty preference, theoretical inclination, and even popular public opinion.

---

[5] Carol Tavris, "The Scientist-Practitioner Gap: Revisiting 'A View from the Bridge' a Decade Later," in *The Science and Pseudoscience of Clinical Psychology*, ed. Scott O. Lilienfeld, Steven J. Lynn, and Jeffrey M. Lohr, 2nd ed. (The Guilford Press, 2015), ix-xx.

Sadly, the vast majority of mental health practitioners blindly run with whatever they were taught in school without asking many questions. After all, students have the right to assume their teachers know what to teach them. Once they graduate, many new therapists don't have a firm foundation based on sound research and solid clinical practice. Yet these budding therapists have the legal authority to diagnose and treat mental disorders, many of which they have not even heard of. The consequences of this can be devastating. Many therapists unwittingly get chewed up and spit out or made a fool of when a narcissistic patient strolls into their office, because they don't have the slightest clue how to help somebody who distorts reality, lacks empathy, and purposely manipulates everyone they encounter so callously and effortlessly. Because of this, many mental health professionals feel insecure about their competency when in the same room with these patients—a vastly different experience from reading about a narcissist in a textbook or hearing about such individuals in a college course on personality theory.

Reading books on narcissism and writing research papers does not qualify anyone to deal with, understand, or effectively treat narcissism. Often, therapists are left with no choice but to fall back on pseudoscience and theories discredited decades ago, thus never truly comprehending all the therapy traps and blunders that come with treating disordered personalities.

Caught wrong-footed, the well-meaning but misinformed therapists —even if they've had success treating traumatized patients—tend to default to a diagnosis of "childhood trauma" as the reason why their narcissistic patients are the way they are. It's almost as if narcissists have gained immunity from treatment from mental health professionals through the misdiagnosis of the root cause of their behavior. Too many therapists don't seem willing to hold narcissists accountable, though they are more than willing to offer them a convenient scapegoat: a bad childhood that wasn't their fault. More often than not, they refer these difficult patients to colleagues with an apology: "I'm having no luck helping this patient. Here, you take a stab at 'em. Oh, and I'm sorry in advance."

If therapists are so misinformed and miseducated, albeit through no fault of their own, how much more misinformed is the general public? A lot. The public has access to unfiltered pseudoscientific ideas via social media and the Internet. The result? Mass confusion.

One of the main goals of this book is to apply empirical research to the examination of the etiology and treatment of narcissistic personality disorder and to avoid theoretical models, popular opinion, and pop psychology. Only through approaching narcissism in this manner will it be possible to dispel the many widely held and *already* discredited myths about narcissism that most people are not aware have already been discredited.

What Is *Not* True about Narcissism and Narcissists

The following represents a list of the most common and most popular misconceptions about the cause of narcissism. Keep in mind, none of the following are true:

- Narcissism is caused by physical abuse in childhood.
- Narcissism is caused by emotional abuse in childhood.
- Narcissism is caused by verbal abuse in childhood.
- Narcissism is caused by sexual abuse in childhood.
- Narcissism is caused by emotional neglect in childhood.

- Narcissism is caused by overindulgence and overvaluation of children by parents.
- Narcissism is caused by the absence of a father in childhood.
- Narcissism is caused by the absence of a mother in childhood.
- Narcissism is caused by conditional love in childhood.
- Narcissism is caused by harsh, cold, or aggressive mothers.
- Narcissism is caused by social and environmental influences.
- Narcissism is caused by bullying in childhood.
- Narcissistic personality disorder is passed down to every subsequent family member and generation.
- Narcissistic parents create narcissistic children.
- Narcissists are the way they are to overcompensate for low self-esteem.
- There are exclusive subtypes of narcissism.
- Narcissists are mentally ill.
- Narcissists get better when their self-esteem increases.
- Narcissists are clinically and legally insane.

- Narcissistic personality traits are determined by parenting.
- All narcissists are highly intelligent master manipulators.
- Children are not narcissistic, but traumatized children grow up to be narcissists at around the age of eighteen or legal adulthood.
- All teenagers exhibit pathologically narcissistic traits due to hormones and brain development.
- Narcissism develops over time and is never present in children who come from warm, loving home environments.
- Everyone is narcissistic to some degree.
- Narcissists are ashamed of themselves.
- Narcissism is a trauma-related disorder, like PTSD.
- Deep down, narcissists just want to be loved.

Again, to be clear: *none* of these prevailing beliefs about narcissism is true, and there is empirical research to prove it. Yet these widely held myths continue to prevail in popular media and culture.

The Price of Being Contrary to Popular Opinion

When I mention that these myths have been discredited by empirical evidence to my colleagues or to the general public, or share valid research on social media in support of this view, the message—and the messenger—is scorned and vilified. I have been accused of being incompetent and irresponsible, and even of promoting dangerous ideas. I've been excoriated on social media for simply sharing hard data, for sharing research that I didn't even conduct personally.

Why is this the case? For one thing, good science often disproves what people *want* to believe. Good science scares, challenges, and upsets people, and for some, it blows their convenient cover.

If you're reading this book, I have to assume that you are at least a little bit curious about why a dedicated therapist would seemingly betray the engrained notions that the vast majority of professionals and people in general have about narcissism. All I ask is that you keep an open mind and see for yourself what the scientific evidence clearly demonstrates.

For those who have suffered devastating consequences as a result of the misinformation that keeps our society in the dark, tarnishes the credibility of academic psychology programs, and limits our well-meaning licensed professionals, this book aims to correct some of the mistakes related to the cause and treatment of narcissism. Only in this way will practitioners and the general public alike have a scientifically informed understanding of the etiology (cause) and intervention (treatment) of narcissistic personality disorder.

Chapter Four
The Origins of Narcissism

The Myth of Narcissus

Narcissism as a descriptive term for personality pathology did *not* originate in the field of psychology the way other descriptive terms like *psychopath* or *borderline* did. Narcissism originally was used as a reference term, its name borrowed from a character in Greek mythology known as Narcissus.

According to the Greek myth, Narcissus was a devastatingly handsome young man who rejected the romantic advances of a nymph named Echo in favor of his own image reflected in a pool of water. In other words, Narcissus rejected communion—whether friendship or romance—in favor of peering into his watery looking glass and obsessively admiring his own handsome face.

The name Narcissus morphed into the descriptive term *narcissism* when several theorists incorporated myth and literature into their interpretations of patient psychopathology. While this comparative approach is not scientific, it is not completely unreasonable because literature often serves to reveal the human condition just as well as if not better than personal and interpersonal experience. The collective works of Shakespeare are timeless examples of how life imitates art sometimes as much as art imitates life. In fact, it was Shakespeare who first coined the term *nature and nurture*.[6]

In the Beginning: "Narcissus-Like"

The first application of the original Greek myth of Narcissus to the field of mental health occurred in 1898, in a case report written by an English-French physician named Havelock Ellis.[7] In addition to his work as a physician, Ellis was very interested in the study of human sexuality, and he used the expression "Narcissus-like" in a case study as a metaphoric comparison concerning young men that Ellis deemed preoccupied with masturbation, suggesting that those preoccupations might be similar to Narcissus's preoccupation with his

---

[6] Matt McGue, "Behavioral Genetics," video file, February 27, 2024, https://www.coursera.org/learn/behavioralgenetics.
[7] Jessica Yakeley, "Current Understanding of Narcissism and Narcissistic Personality Disorder," *BJPsych Advances* 24 (2018): 305-315.

own image in the pool of water at the expense of others.[8]

It is important to stress here that the first reference of *narcissism* in a case study uses a metaphor from Greek mythology, not biology, neurology, or medicine, to explain a particular behavior or tendency in a certain population of patients, namely young boys. Specifically, Ellis suggests a comparison between boys who masturbate on a regular basis, especially excessively, as if this has something to do with excessive self-involvement or investment in oneself similar to Narcissus's preoccupation with his own image in the Greek myth. To be fair, Ellis didn't label these boys narcissists; he said they are "Narcissus-like," which is quite different.

In Ellis's day at the close of the 1800s, narcissism was not a diagnosis, diagnostic term, or label to describe the personality of an individual. The concept was used simply as a reference to suggest self-involvement in certain people, specifically young boys, and Ellis *wondered* if there was

---

[8] Aaron T. Beck and Arthur Freeman, EdDs., *Cognitive Therapy of Personality Disorders* (The Guilford Press, 1990).; Jessica Yakeley, "Current Understanding of Narcissism and Narcissistic Personality Disorder," *BJPsych Advances* 24 (2018): 305-315.

a correlation between excessive self-involvement and autoeroticism. His theory was intended to be of contemplative benefit for those who treated patients. "Narcissus-like" was not intended as an insult to his patients, nor was it intended for general public use. Ellis's ideas were not empirically validated science. His methodology was basically theoretical interpretation, and there were no controlled research studies conducted to validate whether or not Ellis's interpretations and comparisons were accurate. From our modern-day scientific perspective, Ellis's interpretive musings have been proven invalid. It is demonstrably not true that every young boy who masturbates, even a lot, is a self-involved narcissist.

So the Narcissus/narcissism concept has its roots in the interpretation of one theorist from more than a century ago about the self-pleasuring sexual activity of a certain kind of patient. It is also important to note that because Ellis himself was interested in the study of human sexuality, he may have been influenced to theorize from a biased perspective with an emphasis on sexual behavior. This is why subjective interpretations, even those of really smart people, cannot and should not be considered empirically validated science. Too often they involve a confirmation bias on the part of the interpreter.

The Historic Evolution of *Narcissism*

The first person to use the term *narcissism* in a theoretical study was Paul Nacke, a German psychiatrist and criminologist, in the year 1899, and the subject of his study was sexual perversion.[9] We already see a slight theme of bias here with the use of the term *narcissism* in subjective, theoretical case reports beginning as a reference to excessive masturbation and evolving into sexual perversion. Again, these types of reports were based on personal and subjective interpretations of patients or clients and have no empirical evidence to back them up. You'll see why this is so important later.

In 1911, a psychoanalyst named Otto Rank wrote the very first psychoanalytic paper on narcissism. In that paper, he elaborates on the concept by linking it to vanity and self-admiration, specifically in *females*, but the vanity and self-admiration Rank referred to was not considered either exclusively sexual or pathological.[10]

---

[9] Frank R. George and Derek Short, "The Cognitive Neuroscience of Narcissism," *Science Publications* (2017): 1-14.

[10] Frank R. George and Derek Short, "The Cognitive Neuroscience of Narcissism," *Science Publications* (2017): 1-14.; Amanda Green, Rebecca MacLean, and Katherine Charles, "Female Narcissism: Assessment, Aetiology, and Behavioural Manifestations," *Psychological Reports* 125, no. 6 (2022): 2833-2864.; Jessica Yakeley, "Current Understanding of Narcissism and Narcissistic Personality Disorder," *BJPsych Advances* 24 (2018): 305-315.

Enter Freud. In 1914, Sigmund Freud, an Austrian neurologist, popularly known as the founder of psychoanalysis, published a paper entitled "On Narcissism: An Introduction" in which he proposed two different types of narcissism: *primary narcissism* and *secondary narcissism*. According to Freud, primary narcissism is simply a part of normal development and is experienced by everyone. All children are self-centered, Freud suggested. This is not entirely untrue; young children don't yet have the brain wiring to be selfless and altruistic in the same ways adults do. For Freud, primary narcissism has to do with self-preservation and survival instincts and is not pathological by any means.[11]

Secondary narcissism, on the other hand, *is* pathological according to Freud, because it occurs when an individual withdraws their love for others and reinvests that love back into themselves in the way that Narcissus refused Echo's love and instead invested in his own image in the pool of water.[12]

---

[11] Amanda Green, Rebecca MacLean, and Katherine Charles, "Female Narcissism," *Psychological Reports* 125, no. 6 (2022): 2833-2864.
[12] Gary A. Russell, "Narcissism and the Narcissistic Personality Disorder: A Comparison of the Theories of Kernberg and Kohut," *British Journal of Medical Psychology* 58 (1985): 137-148.

Freud, like Otto Rank, also originally considered females more narcissistic than males because he made the false assumption that women were more preoccupied with their appearance than men.[13] Nevertheless, Freud believed that all narcissistic tendencies, including secondary, pathological narcissism, could be "worked through" in psychoanalysis, and to do so was vital for healthy maturation.[14]

While Freud, though demonstrably a genius, is popularly misunderstood by many who haven't read his original work (I recommend reading it), he did make a very telling statement in his essay "On Narcissism: An Introduction" regarding his aim for the field of psychology. In the essay, Freud states: "I try in general to keep psychology clear from everything that is different in nature from it, even biological lines of thought." In other words, for Freud, evidence-based science and theoretical mind science ran down two separate and distinct tracks.

Despite mental health science adopting the health care model in the late 1800s,[15] there

---

[13] Amanda Green, Rebecca MacLean, and Katherine Charles, "Female Narcissism," *Psychological Reports* 125, no. 6 (2022): 2833-2864.
[14] Frank R. George and Derek Short, "The Cognitive Neuroscience of Narcissism," *Science Publications* (2017): 1-14.
[15] Alan Godwin and Gregory W. Lester, *Demystifying Personality Disorders: Clinical Skills for Working with Drama and Manipulation* (PESI Publishing, 2021).

didn't seem to really be an intention or desire on the part of the early pioneers in the field to test their theories and interpretations using scientific methods, which was odd, considering several of the pioneers studied neurology and medicine and were practicing physicians. This is important because it speaks to the bias that has been present in psychological theory since the beginning. The field of mental health care seems to have been shunted aside by its own architects from disciplines such as medicine, biology, and neurology in favor of their own preferences. This perhaps could be one of the main reasons why beliefs about psychology and disorders such as narcissism are conceptualized through the use of pseudoscientific ideas.

The 1960s and 1970s saw a resurgence in interest in narcissism within the field of psychology, specifically for the psychoanalytic theorists Otto Kernberg and Heinz Kohut, whose theories on narcissism are drastically different from one another's despite having come from the same theoretical background and practice.[16]

---

[16] A. Elliot, *Psychoanalytic Theory: An Introduction*, Duke University Press (2000); Gary A. Russell, "Narcissism and the Narcissistic Personality Disorder," *British Journal of Medical Psychology* 58 (1985): 137-148.

Otto Kernberg coined the term *narcissistic personality structure* in the late 1960s[17] and emphasized the aggressive, exploitive, envious, and parasitic features of narcissistic personalities.[18] He further suggested that early childhood experiences with emotionally absent or aggressive mothers forced children to develop a false sense of uniqueness and superiority as a psychological defense against maternal aggression or neglect.[19] According to Kernberg, the grandiose sense of self-importance that pathological narcissists display in adulthood is caused by harsh, uncaring, and aggressive mothers and is all just a front to psychologically defend against secret or hidden rage and envy.[20]

It is especially important to note that according to Kernberg, pathological narcissism can only develop in a child between birth and about

---

[17] Frank R. George and Derek Short, "The Cognitive Neuroscience of Narcissism," *Science Publications* (2017): 1-14.

[18] Frank R. George and Derek Short, "The Cognitive Neuroscience of Narcissism," *Science Publications* (2017): 1-14.; Jessica Yakeley, "Current Understanding of Narcissism and Narcissistic Personality Disorder," *BJPsych Advances* 24 (2018): 305-315.

[19] Frank R. George and Derek Short, "The Cognitive Neuroscience of Narcissism," *Science Publications* (2017): 1-14.; Jessica Yakeley, "Current Understanding of Narcissism and Narcissistic Personality Disorder," *BJPsych Advances* 24 (2018): 305-315.

[20] Gary A. Russell, "Narcissism and the Narcissistic Personality Disorder," *British Journal of Medical Psychology* 58 (1985): 137-148.

eighteen months of age.[21] Kernberg also believed that "pathological narcissism" was different from and unrelated to the "normal narcissism" which he suggested is present in the personality structure of all human beings, as did Freud, as part of the natural course of human psychological development.[22]

Heinz Kohut departs significantly from Freud and Kernberg's views on narcissism.[23] According to Kohut, pathological narcissism can originate anytime between birth and puberty, and he believed narcissism was the result of faulty parenting.[24] Kohut was the first theorist to refer to narcissistic personality as a "disorder" in the late 1960s. The very popular idea that narcissism is a defense to suppress feelings of low self-esteem and inadequacy comes mainly from Kohut's

---

[21] Gerson, M.J. (2021). *Child Abuse and Trauma*. Westlake Village, CA: Institute of Advanced Psychological Studies.; Gary A. Russell, "Narcissism and the Narcissistic Personality Disorder," British Journal of Medical Psychology 58 (1985): 137-148.

[22] Alan Godwin and Gregory W. Lester, *Demystifying Personality Disorders: Clinical Skills for Working with Drama and Manipulation* (PESI Publishing, 2021).

[23] Gary A. Russell, "Narcissism and the Narcissistic Personality Disorder: A Comparison of the Theories of Kernberg and Kohut," *British Journal of Medical Psychology* 58 (1985): 137-148.

[24] M.J. Gerson, *Child Abuse and Trauma*. Westlake Village, CA: Institute of Advanced Psychological Studies (2021).; Gary A. Russell, "Narcissism and the Narcissistic Personality Disorder," *British Journal of Medical Psychology* 58 (1985): 137-148.

work, although Alfred Adler's theory on inferiority complexes, which suggests that people overcompensate for feelings of low self-worth, has also contributed to this widespread idea.[25]

It is quite possible that Kohut may have come to his conclusions because narcissists often show up to treatment presenting as vulnerable or fragile after their lives have decomposed or fallen apart in some major way.[26] This is not because narcissists consistently present with vulnerable traits, but rather because narcissists are typically only motivated to seek help after they experience a crisis or significant interpersonal distress, which makes them *appear* more vulnerable than they actually are in clinical settings.[27]

Even today, overly relying on *how* someone presents in therapy and taking how they "appear" at face value has resulted in skewed understandings and biases among therapists. Though narcissists may prominently present as highly vulnerable

---

[25] Frank R. George and Derek Short, "The Cognitive Neuroscience of Narcissism," *Science Publications* (2017): 1-14.

[26] [26] Joshua D. Miller, Thomas A. Widiger, and W. Keith Campbell, "Vulnerable Narcissism: Commentary for the Special Series 'Narcissistic Personality Disorder – New Perspectives on Diagnosis and Treatment'," *Personality Disorders: Theory, Research, and Treatment* 5, no. 4 (2014): 450-451.

[27] Joshua D. Miller et al., "Vulnerable Narcissism," *Personality Disorders* 5, no. 4 (2014): 450-451.; Jessica Yakeley, "Current Understanding of Narcissism and Narcissistic Personality Disorder," *BJPsych Advances* 24 (2018): 305-315.

in clinical settings, empirical research has shown this not to be the case.[28]

Kohut suggested that narcissistic personalities experience emotional wounds—what he referred to as *narcissistic injuries*—that lead to feelings of emptiness and depression.[29] For Kohut, the narcissist's grandiose exterior is a psychological defense that the narcissist doesn't actually believe about themselves but is used to mask their highly vulnerable interior life.[30] Kohut believed narcissism was "disordered" only when normal narcissism extended into adulthood.[31]

In the late 1970s and early 1980s, American psychologist Theodore Millon incorporated social learning theory into an understanding of the cause and development of narcissism.[32] According to Millon, children are vulnerable to the behaviors they witness occurring in their parents,

[28] Joshua D. Miller et al., "Vulnerable Narcissism," *Personality Disorders* 5, no. 4 (2014): 450-451.; Jessica Yakeley, "Current Understanding of Narcissism and Narcissistic Personality Disorder," *BJPsych Advances* 24 (2018): 305-315.

[29] Jessica Yakeley, "Current Understanding of Narcissism and Narcissistic Personality Disorder," *BJPsych Advances* 24 (2018): 305-315.

[30] Frank R. George and Derek Short, "The Cognitive Neuroscience of Narcissism," *Science Publications* (2017): 1-14.

[31] Gregory W. Lester, *Borderline, Narcissistic, Antisocial, and Histrionic Personality Disorders* (PESI Publishing, 2024).; Gary A. Russell, "Narcissism and the Narcissistic Personality Disorder," *British Journal of Medical Psychology* 58 (1985): 137-148.

[32] Jessica Yakeley, "Current Understanding of Narcissism and Narcissistic Personality Disorder," *BJPsych Advances* 24 (2018): 305-315.

and especially to the behaviors of overindulging parents. Specifically, these observations may cause the observant children to become pathological in their belief that they are more entitled and special than others.[33]

This may come as a shock to many, but these theories on narcissism were completely based on what is known as *anecdotal data*.[34] While some of the identified traits of narcissistic individuals, like grandiosity and arrogance, rage and envy, and beliefs about superiority and entitlement, did later prove to be consistent with and validated by scientific research, no scientific measures or methods were implemented to determine the actual *cause* of narcissism.

Mental health professionals up to this point could only speculate as to what might be causing such a disordered way of thinking, feeling, and behaving in narcissistic individuals. Theories about how bad childhoods compromised normal psychological development prevailed among professionals in the fields of psychiatry and psychology

---

[33] Jessica Yakeley, "Current Understanding of Narcissism and Narcissistic Personality Disorder," *BJPsych Advances* 24 (2018): 305-315.

[34] Gregory W. Lester, *Personality Disorders: Advanced Diagnosis, Treatment & Management* (PESI Publishing, 2018).; Matt McGue, "Behavioral Genetics," video file, February 27, 2024, https://www.coursera.org/learn/behavioralgenetics.

since the beginning, and these explanations were widely accepted without much debate despite being based *solely* on patient self-reports and practitioner interpretations. No wonder so many people today still do not consider psychology to be a real science—can you really blame them?

To be clear, the clinical observations of Freud, Kohut, Kernberg, Millon, and others that describe what a narcissist is actually like to deal with were accurate. That is not the problem. These theorists were accurate in their *descriptions* of pathological narcissism; they just had no empirical evidence to support their *explanations* of pathological narcissism because they did not, and could not, utilize objective, controlled, scientific testing methods to confirm their findings. Had they been able to, their assumptions about the cause of narcissism would have been proven incorrect, and, like good clinicians, they likely would have gone back to the drawing board. It's also entirely possible that some of them did not want to believe there could be any other cause for adult psychological problems besides early childhood experiences with parents or primary caregivers.

Today, empirical research shows that traditional psychological theories, while effective in explaining and dealing with neurotic or internal conflicts and ruptures originating in childhood, do not provide sufficient explanation for the *cause* of narcissism and are not effective *treatments* for narcissism either.

The Fallacy of Anecdotal "Evidence"

Through the 1970s, the data on narcissism was mostly derived from adult patients reporting their childhood experiences to clinicians who took their stories at face value and who were already convinced, based on their training and confirmation biases, that childhood psychological development was the root cause of *all* psychological and emotional problems in adults.[35] The major problem with this method? Anecdotes cannot actually prove nature or nurture simply because anecdotes cannot prove one scientific argument over another.[36] Anecdotes are not scientific data. Anecdotes are stories.

Another big problem with the anecdotal method was that there was no consensus among practitioners on how to effectively treat individuals who presented with narcissism. This was

---

[35] Paris, *Myths of Trauma* (2023).

[36] Matt McGue, "Behavioral Genetics," video file, February 27, 2024, https://www.coursera.org/learn/behavioralgenetics.

especially troublesome to theorists and practitioners because narcissistic patients were simply not getting better in treatment, and some seemed to get even worse.[37] In fact, many became even more dangerous *after* therapy.[38]

Ultimately, treatment methods that proved helpful to patients who were not narcissistic ended up causing more problems than they were solving with narcissistic individuals. The bottom line: mainstream mental health care, from its inception all the way up until the 1980s (late 20th century), was primarily focused on narcissism as viewed through the distorted lens of personal interpretation and pseudoscience. Sadly, this is still the case today.

The Pendulum Shift and the Rise of the *DSM*

In 1980, the third edition of the *Diagnostic and Statistical Manual of Mental Disorders* (*DSM-III*) was published, and for the first time in history, there were clear labels used for mental disorders as well as some diagnostic principles and statistics

---

[37] Gregory W. Lester, *Personality Disorders: Advanced Diagnosis, Treatment & Management* (PESI Publishing, 2018).

[38] Alan Godwin and Gregory W. Lester, *Demystifying Personality Disorders: Clinical Skills for Working with Drama and Manipulation* (PESI Publishing, 2021).

for professionals to utilize in order to estimate the prevalence of the various mental disorders.[39] Moreover, the committee that worked on the updated diagnostic manual chose to eliminate psychoanalytic concepts based solely on subjective patient story-telling and subjective practitioner interpretation.[40] This was quite controversial at the time.

Finally, psychology sought to use experimental, reliable, testable, and controlled scientific methods to accurately diagnose, treat, and explain mental disorders. However, this shift from a reliance on anecdotal data to a more scientific approach presented a number of problems.[41] In a way, the pendulum swung too far in the opposite direction. The result was that psychological models and theoretical concepts were totally dismissed in favor of biological explanations that reduced all psychological problems to predictable symptom clusters that could potentially be treated with the latest prescription medications.[42]

---

[39] Frank R. George and Derek Short, "The Cognitive Neuroscience of Narcissism," *Science Publications* (2017): 1-14.

[40] Gregory W. Lester, *Borderline, Narcissistic, Antisocial, and Histrionic Personality Disorders* (PESI Publishing, 2024). (Lester, 2024)

[41] J. Paris, *Fads and Fallacies in Psychiatry* (Cambridge University Press, 2023).

[42] Paris, *Fads and Fallacies in Psychiatry* (2023).

This pendulum swing is similar to the radical behaviorist movement that began in the 1920s, when, in an effort to distance itself from the eugenics movement, psychology essentially ignored and dismissed any and all science related to genetics, including the already scientifically proven genetic influence on personality in favor of the *blank slate myth*. This long-ago discredited myth sadly still carries a lot of influence in our culture today.[43]

Briefly, the *blank slate myth* is the false notion that babies are born blank slates, as if their minds are blank canvases or blocks of clay that can essentially be shaped or molded into whatever parents or other influential figures desire. Essentially, this myth suggests that there is no such thing as the influence of nature, *only* nurture. The reality is that nature (biology) and nurture (environment) are so dynamically interconnected that you cannot accurately have a conversation about one without the other.[44]

Back to the *DSM*. While this useful and definitive manual was created by practitioners for

---

[43] Matt McGue, "Behavioral Genetics," video file, February 27, 2024, https://www.coursera.org/learn/behavioralgenetics.

[44] Lisa Feldman Barrett, *Seven and a Half Lessons about the Brain* (Mariner Books, 2021).

practitioners, it is still evolving even to this day, and there will likely always be revisions. It is not a scientific masterpiece without limitations. And it's in no way supposed to be considered the Bible of psychiatry and psychology.[45] However, in a way, similar to how the Bible was once reserved for clergy to interpret and then gradually evolved into a book that can be interpreted in countless ways by whoever holds a copy, the *DSM* is now widely used (and misused) as a self-diagnosis manual for anyone who has internet or social media access.

Reluctance to Evolve

Many professionals are not even proficiently trained to utilize the *DSM* manual properly, and this presents both a major public health problem and a major ethical issue in the field of psychology. For example, recent studies have shown that over one-third of severe mental disorder diagnoses rendered by professionals contain errors, and over half of all patients—that's right, *over half*—diagnosed with depressive disorders are misdiagnosed by clinicians.[46] For those who have seen the

---

[45] Paris, *Fads and Fallacies in Psychiatry* (2023).
[46] Bloom, *Mastering Differential Diagnosis with the DSM-5-TR.*

evolution and revisions of the *DSM* from 1980 to its most recent text revision in 2022, things have changed, but in a lot of ways there is still and always will be a long way to go.

All that being said, the officially recognized, categorized, and classified mental disorder known as *narcissistic personality disorder* (NPD) was only introduced forty-four years ago.[47] It generally takes much longer than half a century in any field or discipline to successfully dispel myths and successfully divide pseudoscience from actual science. Therefore, it's really not unusual for both clinicians and the general public to be confused about what's true about narcissism. Even well-regarded academic programs continue to teach what amounts to misinformation. Moreover, many people, including mental health professionals, simply don't like change, and many don't like to admit they may be wrong or they may have been misinformed or miseducated.[48]

The tendency of therapists to steadfastly believe in the validity of clinical methods that *they*

---

[47] Frank R. George and Derek Short, "The Cognitive Neuroscience of Narcissism," *Science Publications* (2017): 1-14.

[48] Carol Tavris, "The Scientist-Practitioner Gap," in The Science and Pseudoscience of Clinical Psychology, ed. Scott O. Lilienfeld, Steven J. Lynn, and Jeffrey M. Lohr, 2nd ed. (The Guilford Press, 2015), ix-xx.

prefer often takes precedence over efficacy and the erroneous assumption that every therapist can draw reliable conclusions about the cause of mental disorders "simply by listening to what people say about their lives."[49] This is a major problem in clinical practice. Therapists who believe that early childhood experiences are the cause of all adult psychological problems, like it or not, are mistaken.[50]

For the record, not only is this bad science but it's bad therapy, and this reluctance to evolve can and has greatly harmed patients and compromised their mental and physical health, not to mention compromised the legitimacy of psychology as a scientific discipline.

In the latest edition of the *DSM* (*DSM-5-TR*), the diagnostic format is more highly organized, and there is an additional section on personality disorders, including narcissism, that focuses less on checklist criteria and more on the frequency and severity of functioning and traits. This inclusion is meant to address the limitations of the categorical checklist model.[51] This admission in

---

[49] Joel Paris, *Myths of Childhood* (Routledge, 2014).

[50] Carol Tavris, "The Scientist-Practitioner Gap," in The Science and Pseudoscience of Clinical Psychology, ed. Scott O. Lilienfeld, Steven J. Lynn, and Jeffrey M. Lohr, 2nd ed. (The Guilford Press, 2015), ix-xx.

[51] American Psychiatric Association. 2022. *Diagnostic and Statistical Manual of Mental Disorders*. 5th ed., text rev.

the *DSM-5-TR* itself of potential shortcomings of checklist-style diagnoses is refreshing because relying on a checklist of criteria alone to describe, explain, diagnose, and treat a complex disorder like narcissistic personality disorder is obviously insufficient.[52]

Separating Personality Disorders from Other Disorders

Why is there so much extra attention focused on personality disorders? Why isn't there a completely separate section in the manual for PTSD, for example? Because personality disorders, contrary to popular belief, are *not* trauma-related disorders or behavioral disorders. Personality disorders, contrary to popular belief, are *not* the result of adverse environments in childhood, or parenting style, or parenting in general, whether the parenting is good or bad. Personality disorders, like narcissism, are characteristic disorders, *not* symptomatic conditions.[53]

---

[52] Aaron L. Pincus, Nicole M. Cain, and Aidan G. Wright, "Narcissistic Grandiosity and Narcissistic Vulnerability in Psychotherapy," *Personality Disorders: Theory, Research, and Treatment* 5, no. 4 (2014): 439-443.

[53] Alan Godwin and Gregory W. Lester, *Demystifying Personality Disorders* (PESI Publishing, 2021).

The point of all of this is that we have come a long way since assuming causal factors of narcissism and basing these causal factors on stories recalled in therapy by patients who are notorious for distorting and misrepresenting the truth. Narcissists intentionally lie; under the guise of help-seeking patients, they dramatize, manipulate, and self-victimize. They have personalities very similar to my maternal grandmother.

Chapter Five
Holding a Mirror Up to Narcissism

Narcissism in Popular Culture vs. Pathological Narcissism

Whatever "normal" narcissism is, or what we might colloquially term *narcissistic personal regard*, it's not really a problem. Tendencies that are reality based and not problematic are not really worth the concern of clinicians or the general public. We all have different personality quirks and tendencies, and as long as they don't interfere with or impair our social or occupational lives, interfere in other important areas of functioning, or impair functioning in others, who really cares? Differences and quirks are eons away from being the same thing as disorders.[54]

So people who don't know how pathological narcissism differs from "healthy" or "normal" narcissism, or who have never had to deal interpersonally with someone who meets the criteria for narcissistic personality disorder for an extended period of time, tend to say things like: "We're all

---

[54] Gregory W. Lester, *Borderline, Narcissistic, Antisocial, and Histrionic Personality Disorders* (PESI Publishing, 2024).

narcissistic to a degree, so stop making a big deal of it," or, "Narcissism is just compensation for low self-esteem," or, "Narcissists are just emotionally stunted children, so give them a break," or, "Narcissists aren't that bad; people are just too sensitive."

Frankly, the people who say these things simply do not know what they are talking about. Not only that, but their erroneous thinking can be insensitive at best and dangerous at worst. On the other hand, the number of people who insist that their spouse, ex-partners, parents, siblings, or in-laws are full-blown malignant narcissists might not have a clear or accurate understanding of what they are saying either. The point is this: we have to be careful not to underestimate *or* overestimate the seriousness of a disorder or the prevalence of a disorder like narcissism. And, like it or not, we do have to leave the actual diagnosis to qualified and competent professionals.

One of the reasons I'm emphasizing this is because narcissism as a diagnostic indicator was never meant to be a pejorative term or a jab at someone who slighted you or wronged you, even if the emotional injury you suffered was devastating. It was never meant to be, and still should not be, a term used as an insult. It was never meant to confuse people. It was never meant to be a topic of popular opinion. It was never meant to be open to personal interpretation. Just like cancer should not be diagnosed outside of the medical discipline and specialty of oncology, narcissism should not be diagnosed outside of psychiatry and psychology.

The term "narcissism" was never intended to be used by the general public the way it has been used in recent years. It was only ever meant to be a descriptive term for clinicians, used by clinicians, to diagnose and clinically treat a disorder that some people meet the criteria for, to help them and those close to them. Period. That's it.

Whatever information exists about narcissism outside the context of clinical research and practice, is, like it or not, very likely based on opinion rather than fact. The reason a term that has fallen prey to pop culture is still used by clinicians and published in clinical literature and diagnostic manuals is really more of a convenience than a desire, because to adopt a linguistic change at this point would be futile and would likely result in even more needless confusion.[55]

---

[55] Gregory W. Lester, *Borderline, Narcissistic, Antisocial, and Histrionic Personality Disorders* (PESI Publishing, 2024).

What Is a Personality Disorder?

As stated before, personality disorders are distinct disorders that were separated from all other mental disorders in diagnostic classification systems because they just didn't fit with any other conditions or syndromes, and patients with these disorders do not respond favorably to traditional treatment interventions.[56] This section will explain why that is the case and what's so distinct about personality disorders like narcissism.

The main distinction between a personality disorder and every other mental health condition is that personality disorders are life-long patterns of continuous disordered functioning *without* periods of normal functioning that affect all or most important areas of personal and interpersonal life, unlike other conditions that have symptoms or episodes that temporarily impair a person's otherwise normal functioning.[57]

Personality disorders don't have symptoms; you can't catch a personality disorder from your environment, and you can't experience a personality

---

[56] Gregory W. Lester, *Personality Disorders: Advanced Diagnosis, Treatment & Management* (PESI Publishing, 2018).
[57] Alan Godwin and Gregory W. Lester, *Demystifying Personality Disorders* (PESI Publishing, 2021).; Paris, *Fads and Fallacies in Psychiatry* (2023).

disorder "episode" like people with major depression or bipolar disorder can, for example.[58] Symptoms come and go. Personality disorders don't. Personality disorders are permanent fixtures on a person like a fingerprint is on a finger.

As stated before, while the *DSM* is not the end-all-be-all authority on every aspect of psychopathology, it is a practical manual that makes it possible to understand and treat mental disorders up to this point in history,[59] so it is well worth turning to for guidance. The *DSM-5-TR* (2022) lists general criteria for personality disorders, criteria that all ten recognized personality disorders, including narcissism, have in common. (See table 1.)

---

[58] Gregory W. Lester, *Borderline, Narcissistic, Antisocial, and Histrionic Personality Disorders* (PESI Publishing, 2024).
[59] Paris, *Fads and Fallacies in Psychiatry* (2023).

Table 1. General personality disorder criteria[60]

A. An enduring pattern of inner experience and behavior that deviates markedly from the expectations of the individual's culture. This pattern is manifested in two (or more) of the following areas:

1. Cognition (i.e., ways of perceiving and interpreting self, other people, and events).
2. Affectivity (i.e., the range, intensity, lability, and appropriateness of emotional response).
3. Interpersonal functioning.
4. Impulse control.

B. The enduring pattern is inflexible and pervasive across a broad range of personal and social situations.

C. The enduring pattern leads to clinically significant distress or impairment in social, occupational, or other important areas of functioning.

D. The pattern is stable and of long duration, and its onset can be traced back at least to adolescence or early adulthood.

E. The enduring pattern is not better explained as a manifestation or consequence of another mental disorder.

---

[60] APA. 2022. *Diagnostic and Statistical Manual of Mental Disorders.* 5th ed., text rev.

F. The enduring pattern is not attributable to the physiological effects of a substance (e.g., a drug of abuse, a medication) or another medical condition (e.g., head trauma).

Reprinted with permission from the *Diagnostic and Statistical Manual of Mental Disorders*, Fifth Edition, Text Revision (Copyright 2022). American Psychiatric Association.

Simply put, an individual with a personality disorder like narcissism is an individual who thinks, feels, and behaves in ways that are so stubborn and uncompromising, it makes their character extremely inappropriate and problematic.[61]

How Is Narcissistic Personality Disorder Officially Diagnosed?

Not everyone who behaves in a "narcissistic" fashion is a pathological narcissist or has a disordered personality. There are many variables and factors to consider when formulating an official diagnosis like narcissistic personality disorder. Often, what appears to be narcissism can actually be

---

[61] Alan Godwin and Gregory W. Lester, *Demystifying Personality Disorders* (PESI Publishing, 2021).

explained by overlapping features of other mental conditions or disorders, or can be from the effects of drug or alcohol abuse, the result of a serious medical condition that impairs brain and nervous system functioning, the result of head trauma or a traumatic brain injury, a normal developmental stage of life, such as adolescence, or the result of an individual's sociocultural environment.[62]

All of the above possibilities need to be considered and ruled out *before* a diagnosis of narcissistic personality disorder can officially be rendered. In addition, a diagnosing clinician has to take into consideration a person's full history and not rely solely on a few anecdotes about a person's behaviors from others or from the person in question themselves.

Narcissism is clinically diagnosed based on several factors: *personality functioning*, meaning the level of disturbance in oneself and in one's relationships; *personality traits*, meaning the variation, deficiency, and excess of personality traits; and the *persistent inflexibility of behaviors regardless of consequences*.[63]

---

[62] APA. 2022. *Diagnostic and Statistical Manual of Mental Disorders*. 5th ed., text rev.
[63] APA. 2022. *Diagnostic and Statistical Manual of Mental Disorders*. 5th ed., text rev.

To be fair, the *DSM* offers diagnostic criteria, but it does not offer the complete picture of narcissism by any means, nor does it claim to.[64] Pathological narcissism always reveals a persistent pattern in personality that includes grandiosity, a need for admiration, and a lack of empathy. Currently, the criteria that need to be met for narcissistic personality disorder to be diagnosed officially are listed in the *DSM-5-TR*. (See table 2.)

---

[64] APA. 2022. *Diagnostic and Statistical Manual of Mental Disorders*. 5th ed., text rev.; Aaron L. Pincus et al. "Narcissistic Grandiosity and Narcissistic Vulnerability in Psychotherapy," *Personality Disorders* 5, no. 4 (2014): 439-443.

Table 2. Criteria for narcissistic personality disorder (*DSM-5-TR*, 2022)

A pervasive pattern of grandiosity (in fantasy or behavior), need for admiration, and lack of empathy, beginning by early adulthood and present in a variety of contexts, as indicated by five (or more) of the following:

1. Has a grandiose sense of self-importance (e.g., exaggerates achievements and talents, expects to be recognized as superior without commensurate achievements).
2. Is preoccupied with fantasies of unlimited success, power, brilliance, beauty, or ideal love.
3. Believes that he or she is "special" and unique and can only be understood by, or should associate with, other special or high-status people (or institutions).
4. Requires excessive admiration.
5. Has a sense of entitlement (i.e., unreasonable expectations of especially favorable treatment or automatic compliance with his or her expectations).
6. Is interpersonally exploitative (i.e., takes advantage of others to achieve his or her own ends).
7. Lacks empathy: is unwilling to recognize or identify with the feelings and needs of others.
8. Is often envious of others or believes that others are envious of him or her.
9. Shows arrogant, haughty behaviors or attitudes.

Reprinted with permission from the Diagnostic and Statistical Manual of Mental Disorders, Fifth Edition, Text Revision (Copyright 2022). American Psychiatric Association.

Diagnostic criteria aside, what is it like to interact with a narcissist? Narcissists tend to cause a knee-jerk response in those who have had prolonged contact with them. "Oh God," accompanied by an eye roll and a heavy sigh is a common reaction. Narcissists cause upset, confusion, frustration, and conflict wherever they go,[65] although not at first. Who they really are (and aren't) at their core becomes obvious only after the thin veneer of their initial—often charming—camouflage wears off.

How a Narcissist Thinks, Feels, Behaves, and Perceives Reality

Narcissists are envious of others, but they mask their envy as respect or admiration in an attempt to obtain what they need from someone who they in fact envy. They can feign being very accommodating and appreciative if they believe you can serve them in some way.[66]

Narcissists are deficient in the capacity to feel gratitude toward others. They are convinced

---

[65] Gregory W. Lester, *Personality Disorders: Advanced Diagnosis, Treatment & Management* (PESI Publishing, 2018).
[66] Donald J. Robinson, *Disordered Personalities*, 3rd ed. (Rapid Psychler Press, 2005).

they don't need others, so they perceive needing others as a sign of weakness. Due to their empathy deficits, they have little to no capacity to care for others, to share with others, or to love others.[67]

Narcissists believe that by nature, they are better and superior human beings.[68] They hold high views of themselves even if those views are in stark contrast with their reality.[69] They like to be in charge and believe themselves to be competent even when they aren't.[70] They overestimate their attractiveness even when others don't find them attractive or appealing.[71] They believe themselves to be brilliant even when their IQ scores prove otherwise.[72] They exaggerate their talents

[67] Donald J. Robinson, *Disordered Personalities*, 3rd ed. (Rapid Psychler Press, 2005).

[68] Eddie Brummelman et al., "What Separates Narcissism from Self-Esteem? A Social-Cognitive Perspective," in *Handbook of Trait Narcissism: Key Advances, Research Methods, and Controversies*, ed. Anthony D. Hermann, Amy B. Brunell, and Joshua D. Foster (Springer International Publishing, 2018), 47-55.

[69] Eddie Brummelman et al., "What Separates Narcissism from Self-Esteem?" in *Handbook of Trait Narcissism*, ed. Anthony D. Hermann, Amy B. Brunell, and Joshua D. Foster (Springer, 2018), 47-55.

[70] Eddie Brummelman et al., "What Separates Narcissism from Self-Esteem?" in *Handbook of Trait Narcissism*, ed. Anthony D. Hermann, Amy B. Brunell, and Joshua D. Foster (Springer, 2018), 47-55.

[71] Eddie Brummelman et al., "What Separates Narcissism from Self-Esteem?" in *Handbook of Trait Narcissism*, ed. Anthony D. Hermann, Amy B. Brunell, and Joshua D. Foster (Springer, 2018), 47-55.

[72] Eddie Brummelman et al., "What Separates Narcissism from Self-Esteem?" in *Handbook of Trait Narcissism*, ed. Anthony D. Hermann, Amy B. Brunell, and Joshua D. Foster (Springer, 2018), 47-55.

and achievements even when they don't have anything substantial to show for them. Because of all of this, narcissists look down on and dehumanize others.[73]

Narcissists cannot tolerate when things aren't perfect, and because nothing is ever perfect in reality, they don't believe that anything or anyone is good enough for them. This is why they don't value fidelity and loyalty and have no problem betraying others.[74]

Narcissists psychologically block out other people when they are talking to them and relating to them, as if they are really just talking to a mirror. When the mirror doesn't reflect back exactly what they want, which it never can in reality, they become irritated or enraged. This is why they cannot tolerate being challenged, disagreed with, or insufficiently admired.[75]

Narcissists are emotionally impulsive and can lash out when they experience negative emotions, but not always overtly. They project their negative emotions onto others, which causes

---

[73] Eddie Brummelman et al., "What Separates Narcissism from Self-Esteem?" in *Handbook of Trait Narcissism*, ed. Anthony D. Hermann, Amy B. Brunell, and Joshua D. Foster (Springer, 2018), 47-55.

[74] Donald J. Robinson, *Disordered Personalities*, 3rd ed. (Rapid Psychler Press, 2005).

[75] Donald J. Robinson, *Disordered Personalities*, 3rd ed. (Rapid Psychler Press, 2005).

the recipient of their lash-outs to experience the narcissist's negativity almost as a physical blow. They also possess an uncanny ability to extract good feelings from those who get caught in their drama.[76]

Narcissists don't relate to people unless the person they relate to serves a particular purpose for them. They are highly insufficient at collaborating and cooperating with others. Narcissists only use people for their own one-sided purposes and needs.[77]

Narcissists are responsibility avoiders. They deny their part in any conflict or problem, they lie constantly, both deliberately and automatically, and they prematurely quit and give up on tasks, jobs, relationships, and responsibilities that don't perfectly mirror their idealized views or sufficiently provide for their selfish needs.[78]

Narcissists are exploitative, controlling, demanding, oppositional, manipulative, parasitic, and threatening. They may use blackmail and other

---

[76] Gregory W. Lester, *Borderline, Narcissistic, Antisocial, and Histrionic Personality Disorders* (PESI Publishing, 2024).

[77] Alan Godwin and Gregory W. Lester, *Demystifying Personality Disorders* (PESI Publishing, 2021).

[78] Gregory W. Lester, *Personality Disorders: Advanced Diagnosis, Treatment & Management* (PESI Publishing, 2018).

coercive tactics to force people to submit to their needs and desires, and they have a high potential for verbal and physical violence when they don't get what they want. Their sole purpose for entering into relationships with others is to ensure that some desirable result happens for them and them alone.[79]

Narcissists are thoughtless, insensitive, and uncaring toward others. They feel entitled to treat people any way they want and discard them when they are no longer sufficiently providing for their needs. They truly believe they are special and superior to everyone else, and because of this, they make exceptions for themselves related to social and relational standards and rules that everyone else willingly accepts and abides by. They do not value or believe in equality.[80]

Narcissists are limited to dichotomous thinking because they are rigid and inflexible thinkers. Thoughts are all or nothing, good or bad, and right or wrong, with no gray area or middle ground. With narcissists, this dichotomy always works in *their* favor. For example, the narcissist is always right and

---

[79] George Simon Jr., *In Sheep's Clothing: Understanding and Dealing with Manipulative People*, rev. ed. (Parkhurst Brothers Publishers, 2010).
[80] Alan Godwin and Gregory W. Lester, *Demystifying Personality Disorders* (PESI Publishing, 2021).

always innocent, and everyone else is always wrong and always to blame.

Contrary to popular belief, narcissists actually feel a broad range of emotions; however, their impulsivity, heightened arousal, and lack of self-regulating capacity result in dramatic, ineffective escalations rather than normal expressions of feelings and internal states.[81]

The Impact Narcissists Have on Those Around Them

People within a narcissist's orbit describe them in the following ways: selfish, lacking in empathy, high maintenance, difficult, adolescent or childish, toxic, two-faced, full of themselves, always the victim, back-stabber, manipulative, callous, deceitful, angry, insensitive, mean, arrogant, scary, unfair, demeaning, stubborn, ridiculous, spoiled brat, crazy making, degrading, presumptuous, overly competitive, envious, cruel, jerk, baby, unpredictable, and evil, among other things.

People within a narcissist's orbit describe their reaction to interacting with them in the

[81] Gregory W. Lester, *Borderline, Narcissistic, Antisocial, and Histrionic Personality Disorders* (PESI Publishing, 2024).

following ways: with irritability, frustration, fear, dread, automatic appeasing, pacifying, accommodating responses, avoidance, resistance, passive-aggression, covert aggression, resentment, disdain, disgust, a desperate need to try to get one's perspective validated, a desire for retaliation, a desperate need to get the narcissist to accept accountability, a desperate need to defend oneself, and a desperate desire to get the narcissist to tell the truth when caught in a lie, among other things.[82]

Should I Stay or Should I Go?

It can be extremely difficult to make the decision to try to work on a relationship with a narcissistic individual or to choose to end the relationship. This is especially difficult when it comes to issues such as marriage, children, and financial dependency.

If a narcissistic individual has the desire to change, seeks treatment, and successfully develops some flexibility in their trait expression, develops some ability to self-correct and take accountability

---

[82] Gregory W. Lester, *Borderline, Narcissistic, Antisocial, and Histrionic Personality Disorders* (PESI Publishing, 2024).

along with some ability to collaborate and problem solve without drama, there is hope.[83] However, this requires a lot of work on their part and a lot of patience on yours.

Ultimately, you have to make the decision, and here's the question you need to ask yourself when deciding what to do about a relationship with someone who meets the criteria for a personality disorder like narcissism: excluding reasonable levels of life stress and conflict, does this person consistently bring more joy to my life or more suffering? If the answer is the former and the narcissist in question is willing to put in the work, then you have your answer. If the narcissist is the source of greater suffering, you likewise have your answer.

---

[83] Alan Godwin and Gregory W. Lester, *Demystifying Personality Disorders* (PESI Publishing, 2021).

Chapter Six
The Narcissistic Cycle of Abuse

Getting caught in a narcissist's cycle of abuse can easily happen if you don't know what to watch out for. While interacting with a narcissist can feel very destabilizing, there are some predictable phases of interaction you can be on the lookout for. In other words, armed with some factual information, you can become skilled at recognizing the cycle of abuse before you get hooked and avoid the soul-draining cycle altogether, or, if you are already in a narcissist's orbit, you can begin to recognize it for what it is and take steps to get out of it for good.

The Narcissist in Phases

A narcissist's influence and control over others doesn't happen immediately upon contact; it happens gradually over time in much the same way the proverbial frog in the pot of water on the stove doesn't realize the water is slowly heating up to a boil. The best way to look at a narcissist's expanding zone of influence over another is to break it down into phases:

- **Phase One:** The narcissist projects their own ideal image onto another person they find desirable, and they idealize this person, but only in image and only superficially, not in reality.

- **Phase Two:** The narcissist begins to exploit this person, taking everything that is of emotional value from them, without giving anything in return.
- **Phase Three:** The narcissist eventually depletes the other person as a source of gratification.
- **Phase Four:** The narcissist then devalues the depleted person that originally was used as their source of gratification.
- **Phase Five:** The narcissist begins to reject the depleted person, eventually discarding them altogether as a bad object.
- **Phase Six:** The narcissist enters into full-blown reactive mode, shifting between internal states of boredom, emptiness, rage, and distress.

Then, the phase cycle repeats itself when the narcissist once again projects their ideal image onto the next person that they find desirable and begins to target them.[84]

Narcissism and the Self-Esteem Debate

The role self-esteem plays in narcissism seems to be one of the most popular and heated debates in the current day. The main controversy surrounds this: do narcissists display grandiosity and arrogance in order to compensate for hidden or suppressed feelings of low self-esteem and low self-worth? Are they really just putting up a front to defend against their deep-seated feelings of inadequacy and inferiority that resulted from empathic failures and other parenting ruptures in early childhood?

The answer to both questions above is a simple "no." This is where we need to bring science into the debate. The latest research shows us this: the type of narcissism commonly referred to as *vulnerable narcissism* is comprised of non-pathological

---

[84] Otto F. Kernberg, *Internal World and External Reality* (Aronson, 1980).; Donald J. Robinson, *Disordered Personalities*, 3rd ed. (Rapid Psychler Press, 2005).; Dragan M. Svrakic, "The Functional Dynamics of the Narcissistic Personality," *American Journal of Psychotherapy* 44, no. 2 (1990): 189-203.

traits and varying levels of neurotic traits.[85] This means that at best, a vulnerable narcissist is neurotic, and neurosis is not pathology.

We are *all* neurotic to a degree. Anyone who has ever experienced an internal conflict is considered to be neurotic. And we all have. To put this everyday/everyone type of narcissism in the same category as the grandiose, pathological narcissism that describes narcissistic personality disorder is erroneous and needlessly adds to the confusion about narcissism. Moreover, recent research shows that narcissism and self-esteem are so weakly related that it's not really even worth the debate.[86]

There are so many other more important and central features of narcissism to try to understand that spending time discussing the negligible connection between narcissism and self-esteem is hardly worth it. Moreover, doing so won't help clinicians treat narcissists appropriately because there

---

[85] Brendan Weiss and Joshua D. Miller, "Distinguishing between Grandiose Narcissism, Vulnerable Narcissism, and Narcissistic Personality Disorder," in *Handbook of Trait Narcissism: Key Advances, Research Methods, and Controversies*, ed. Anthony D. Hermann, Amy B. Brunell, and Joshua D. Foster (Springer International Publishing, 2018), 3-13.

[86] Eddie Brummelman et al., "What Separates Narcissism from Self-Esteem?" in *Handbook of Trait Narcissism*, ed. Anthony D. Hermann, Amy B. Brunell, and Joshua D. Foster (Springer, 2018), 47-55.

is no evidence whatsoever that increased self-esteem in a narcissist removes their narcissism.

It's also a false assumption that just because someone's grandiosity or sense of entitlement is closeted or covert means they are suffering from a sense of low self-worth and esteem.[87] For example, someone can be exceedingly high on trait introversion—or the preference to focus on inner thoughts and ideas rather than what's occurring externally—but this doesn't correlate to low self-worth or low self-esteem. Introversion has absolutely nothing to do with being shy or insecure.

There is also a difference between what is deemed *covert narcissism*, *covert aggression*, and *covert manipulation*.[88] Covert narcissists are not compensating for low self-esteem just because they aren't the life of the party. Research has shown that pathological grandiosity is present in both covert *and* overt narcissists and is actually a corollary of *high* self-esteem in both overt *and* covert narcissists.[89] Narcissists who are either extroverted or

---

[87] Joshua D. Miller et al., "Vulnerable Narcissism," *Personality Disorders* 5, no. 4 (2014): 450-451.

[88] George Simon Jr., *In Sheep's Clothing*, rev. ed. (Parkhurst Brothers, 2010).

[89] Eddie Brummelman et al., "What Separates Narcissism from Self-Esteem?" in *Handbook of Trait Narcissism*, ed. Anthony D. Hermann, Amy B. Brunell, and Joshua D. Foster (Springer, 2018), 47-55.

introverted, who are either outspoken or withdrawn, hold themselves in their own version of high esteem based on their pervasive sense of arrogance and grandiosity.

The bottom line is this: there is no such thing as low self-esteem in pathological narcissism.

What about *fragile* self-esteem? That's another story. Once their cover is blown, a narcissist can deflate, leading to distress and impulsive acts, even including suicide, but this is not due to low self-esteem or low self-worth. Rather, suicide in narcissists is most often the result of the inability to tolerate humiliation or ideal image failure.[90]

When we understand variations in personality trait excess and personality trait deficiency in each individual, we see personality trait flexibility operating on a continuum. It then becomes abundantly clear how absurd it is to continue to associate narcissistic personalities with low self-esteem and low self-worth. *If* a narcissist is experiencing low self-worth, then there *must* be a co-occurring or comorbid condition, such as depression. And it would be the *depression* or other comorbid

---

[90] Alan Godwin and Gregory W. Lester, *Demystifying Personality Disorders* (PESI Publishing, 2021).

condition that is causing them to feel the low self-worth, *not* the narcissism.

Comorbid conditions are completely separate from the personality pathology of the narcissist. Comorbid conditions occur simultaneously with one another but are not related to or caused by one another. It's the equivalent of having a broken arm and asthma at the same time. They just happen to be occurring simultaneously, but they are completely unrelated to one another.

Back to self-esteem. The problem with self-esteem in the narcissist is not that they have low self-esteem, but that they have very high self-esteem even when it isn't deserved or earned.[91] They aren't faking their opinion of themselves. They *truly* believe they are superior. And people *feel* what they truly *believe* about themselves because both their brain and body make them.[92]

Belief and feeling are not two separately occurring entities. It's not possible to *believe* you are the best and *feel* like you are the worst at the

---

[91] Gregory W. Lester, *Borderline, Narcissistic, Antisocial, and Histrionic Personality Disorders* (PESI Publishing, 2024).
[92] Lisa Feldman Barrett, *Seven and a Half Lessons about the Brain* (Mariner Books, 2021).

same time. That's not how the brain works. That's not how the body works. We feel what we believe. Even introverted narcissists believe they are superior, so therefore, they *feel* superior. You cannot believe that you are special and unique and better than everybody else, and at the same time feel inferior to everyone else.

I'm not sure how this idea of the separation of belief and feeling got so popular or how it has become such a firmly held opinion, because it's just impossible. It's as impossible as a highly anxious person feeling panic while believing they are perfectly safe. A highly anxious person cannot unconsciously be experiencing extreme self-confidence, self-assuredness, and fearlessness right in the midst of a full-blown panic attack. It doesn't make any sense because it's not possible.

It might be that because narcissists lie so often and so naturally and effortlessly that people assume they are lying about themselves as far as how great they feel about themselves. They're not lying in that regard. It's a cognitive error they make about how superior they are to others, but it's a cognitive error that they truly believe in 100 percent.

Understanding What Is Meant by *Vulnerable Narcissist*

In clinical settings, narcissists rarely, if ever, come to treatment stating they are narcissists and need help with their narcissism. They usually come to treatment after a crisis, or after their lives have decomposed in some way. For example, after an extramarital affair or some major life stressor such as the loss of a job, they arrive at a therapist's doorway and report that they are there because they need help dealing with their anxiety, depression, or stress.[93] Narcissism is not something they acknowledge.

Because of this, narcissism is often overlooked and missed completely by many professionals as a viable diagnosis because the patient often presents as vulnerable, depressed, or stressed out.[94] In fact, this oversight is predominantly where the dimension or subtype of *vulnerable narcissism* originates—with patients presenting with

---

[93] Jessica Yakeley, "Current Understanding of Narcissism and Narcissistic Personality Disorder," *BJPsych Advances* 24 (2018): 305-315.
[94] Joshua D. Miller et al., "Vulnerable Narcissism," *Personality Disorders* 5, no. 4 (2014): 450-451.; Jessica Yakeley, "Current Understanding of Narcissism and Narcissistic Personality Disorder," *BJPsych Advances* 24 (2018): 305-315.

vulnerability in treatment after a crisis,[95] and with therapists mistaking their temporary vulnerable presentation as their baseline level of functioning.

Empirical research does not consistently account for vulnerability as a central or even necessary feature of pathological narcissism.[96] Contrary to popular belief, personality disorder researchers and experts do not consider traits such as vulnerability, insecurity, or self-consciousness to be related to pathological narcissism.[97] Those traits are seen more as relating to neuroticism, which is not a pathological component of narcissism.[98]

As outlined previously, the traits that experts closely relate to pathological narcissism are grandiosity, self-centeredness, entitlement, manipulativeness, inequality, and callousness, among others.[99] To be clear, this doesn't mean that narcissists

---

[95] Joshua D. Miller et al., "Vulnerable Narcissism," *Personality Disorders* 5, no. 4 (2014): 450-451.

[96] Joshua D. Miller et al., "Vulnerable Narcissism," *Personality Disorders* 5, no. 4 (2014): 450-451.

[97] Joshua D. Miller et al., "Vulnerable Narcissism," *Personality Disorders* 5, no. 4 (2014): 450-451.

[98] E. N. Aslinger, S. P. Lane, D. R. Lynam, and T. J. Trull, "The Influence of Narcissistic Vulnerability and Grandiosity on Momentary Hostility Leading Up to and Following Interpersonal Rejection," *Personality Disorders: Theory, Research, and Treatment* 13, no. 3 (2022): 199-209.

[99] Gregory W. Lester, *Personality Disorders: Advanced Diagnosis, Treatment & Management* (PESI Publishing, 2018). Joshua D. Miller et al., "Vulnerable Narcissism," *Personality Disorders* 5, no. 4 (2014): 450-451.

don't experience vulnerability. They typically will become vulnerable after their lives begin to fall apart, and narcissists do experience distress. They are, in fact, human beings, even if they don't act like or treat others like a human being should, and they do have feelings. It's just that the vulnerability and distress they experience is not related to self-esteem, self-consciousness, or insecurity; it's related to humiliation resulting from their image shattering or their control being taken away. They may appear vulnerable in therapy as a result of this, and many therapists mistake this as the way the narcissist operates in daily life—like a human being vulnerable to self-consciousness or insecurity—but that's not it at all.

In sum, the subtype of *vulnerable narcissism* is recognized by some researchers and discounted by others. When it comes to vulnerable narcissism, a consensus among experts about its existence and nature has not been fully reached.[100] I am inclined to consider what is referred to as vulnerable narcissism to be a benign, neurotic type of normal narcissism that does not characterize problematic, pathological narcissism.

---

[100] Aslinger et al., "Influence of Narcissistic Vulnerability and Grandiosity" (2021).

In fact, there is a universally accepted consensus among experts and researchers that grandiose features of pathological narcissism are central features and are required to diagnose narcissistic personality disorder. Vulnerable features are not considered essential and therefore are not required to render a formal diagnosis of narcissistic personality disorder.[101] Neurotic narcissistic tendencies develop through social learning and are malleable. They can be reversed and unlearned if the environment changes and new learning takes place. Basically, neurotic narcissism is benign and is not a problem.

---

[101] Joshua D. Miller et al., "Vulnerable Narcissism," *Personality Disorders* 5, no. 4 (2014): 450-451.

Chapter Seven
Cause and Effect

The Myth of Exclusive Subtypes of Narcissism

Clinical research does not distinguish overt and covert narcissism as two distinct subtypes, although pop psychology literature does.[102] There is no such thing as a grandiose narcissist who is only overt or a vulnerable narcissist who is only covert.[103] It is more accurate to say that pathological narcissism can include both overt and covert expressions of trait deficiencies, excesses, and troublesome behaviors. Researchers have argued that there are not any purely distinct subtypes of narcissism, but narcissistic individuals have varying degrees of trait expression and behavior.[104] The central features of narcissism are traits that are deficient or excessive and cause high levels of social impairment.[105]

---

[102] Aslinger et al., "Influence of Narcissistic Vulnerability and Grandiosity" (2021).
[103] Aaron L. Pincus et al. "Narcissistic Grandiosity and Narcissistic Vulnerability in Psychotherapy," *Personality Disorders* 5, no. 4 (2014): 439-443.
[104] Aslinger et al., "Influence of Narcissistic Vulnerability and Grandiosity" (2021).
[105] Alan Godwin and Gregory W. Lester, *Demystifying Personality Disorders* (PESI Publishing, 2021).

So, while it may not be a popular media opinion, it is a fact that purely distinct subtypes of narcissism do not exist in reality. This is an enduring misconception that many social media—entrenched "gurus" have embraced wholeheartedly and, perhaps worse, have developed large followings with, making a fortune from perpetuating false information and fake ideas.

Research Marches On

When it comes to what causes health conditions of all kinds, advances in research and technology have brought us a long way from personal interpretations. Because psychological impairments are directly related to the brain and nervous system, and because personality and temperament have direct links to genetic factors, it was only a matter of time before research in neuroscience and genetic studies would become useful in trying to understand what actually *causes* narcissism and make clear distinctions between *cause* and social-environmental *risk factors*.[106]

---

[106] J. Paris, *Treatment of Borderline Personality Disorder: A Guide to Evidence-Based Practice*, 2nd ed. (The Guilford Press, 2020).

Even though the general public might not be privy to what is currently being compiled in research databases and tested in laboratories, this does not mean that new research is not being conducted or that new findings won't discredit previous research that was at one time believed to be sound, accurate, and representative of all the answers we'd ever need. Good science is always willing to prove its previous findings wrong if those findings no longer hold up. If the field of psychology is going to be taken seriously and ultimately be considered a credible science, we have to leave the research findings to the research experts, not to the popular psychology writers and media influencers.

Enter Genetics

Genetic findings have concluded that narcissism is highly heritable.[107] Family, twin, and adoption

---

[107] Laura A. Baker, "The Biology of Relationships: What Behavioral Genetics Tells Us about Interactions among Family Members," *De Paul Law Review* 56, no. 3 (2007): 837-846.; S. L. Brown and J. R. Young, *Women Who Love Psychopaths: Inside the Relationships of Inevitable Harm with Psychopaths, Sociopaths, and Narcissists*, 3rd ed. (Mask Publishing, 2018).; Frederick L. Coolidge, Leslee L. Thede, and Kerry L. Jang, "Heritability of Personality Disorders in Childhood: A Preliminary Investigation," *Journal of Personality Disorders* 15, no. 1 (2001): 33-40.; Robert O. Friedel, Christine Schmahl, and Marijn Distel, "The Neurobiological Basis of Borderline Personality Disorder," in *Neurobiology of Personality Disorders*, ed. Christine Schmahl et al. (Oxford University Press, 2018), 279-317.; Robert D. Hare, *Hare PCL-R*, 2nd ed. (Multi-Health Systems,

studies actually concluded this as far back as the 1960s. Behavioral and molecular genetic studies have proven this. Neuroscience studies have proven this. Neurobiological studies have proven this.

Inc., 2003).; W. John Livesley and Kerry L. Jang, "The Behavioral Genetics of Personality Disorder," *Annual Review of Clinical Psychology* 4 (2008): 247-274.; Y. L. Luo and H. Cai, "The Etiology of Narcissism: A Review of Behavioral Genetic Studies," in *Handbook of Trait Narcissism: Key Advances, Research Methods, and Controversies*, ed. Anthony D. Hermann, Amy B. Brunell, and Joshua D. Foster (Springer International Publishing, 2018), 149-156.; Guo Ma et al., "Genetic and Neuroimaging Features of Personality Disorders: State of the Art," *Neuroscience Bulletin* 32, no. 3 (2016): 286-306.; Siddhartha Mukherjee, *The Gene: An Intimate History* (Scribner, 2017; I. Nenadic, C. Lorenz, and C. Gaser, "Narcissistic Personality Traits and Prefrontal Brain Structure," *Scientific Reports* 11 (2021), https://doi.org/10.1038/s41598-021-94920-z.; J. Paris, *Treatment of Borderline Personality Disorder: A Guide to Evidence-Based Practice* (2020).; L. A. Pervin, *The Science of Personality*, 2nd ed. (Oxford University Press, 2003).; Ted Reichborn-Kjennerud and Kenneth S. Kendler, "Genetics of Personality Disorders," in *Neurobiology of Personality Disorders*, ed. Christine Schmahl et al. (Oxford University Press, 2018), 57-73.; B. W. Roberts et al., "The Power of Personality: The Comparative Validity of Personality Traits, Socioeconomic Status, and Cognitive Ability for Predicting Important Life Outcomes," *Perspectives in Psychological Science* 2, no. 4 (2007): 313-345.; Sandra Sanchez-Roige et al., "The Genetics of Human Personality," *Genes, Brain, & Behavior* 17, no. 3 (2018), https://doi.org/10.1111/gbb.12439.; George K. Simon Jr., *Character Disturbance: The Phenomenon of Our Age* (Parkhurst Brothers Publishers, 2011).; S. C. South and N. J. DeYoung, "Behavior Genetics of Personality Disorders: Informing Classification and Conceptualization in DSM-5," *Personality Disorders: Theory, Research, and Treatment* 4, no. 3 (2013): 270-283.; Svenn Torgersen et al., "A Twin Study of Personality Disorders," *Comprehensive Psychiatry* 41, no. 6 (2000): 416-425.; Z. E. Wright, S. Pahlen, and R. F. Krueger, "Genetic and Environmental Influences on Diagnostic and Statistical Manual of Mental Disorders-Fifth Edition (DSM-5) Maladaptive Personality Traits and Their Connections with Normative Personality Traits," *Journal of Abnormal Psychology* 126, no. 4 (2017): 416-428.; Mark Zimmerman, "Overview of Personality Disorders," September 2023, accessed March 31, 2024, https://www.merckmanuals.com/professional/psychiatric-disorders/personality-disorders/overview-of-personality-disorders.; I. Zwir et al., "Uncovering the Complex Genetics of Human Character," *Molecular Psychiatry* 25 (2020): 2295-2312.

Evolutionary studies have proven this. Neuroimaging results have shown this. It's not a theory. It's not up for debate. It's not my opinion, and it's not anyone else's opinion. It is hard data, and it is irrefutable.

Taking the multitude of twin studies as one example of the evidence, monozygotic (commonly known as *identical*) twins share 100 percent of their DNA, while dizygotic (commonly known as *fraternal*) twins only share 50 percent. Twin studies that are controlled and that include identical twins raised apart, raised together, and raised in adverse environments as well as nurturing environments are some of the most evident indicators that pathological narcissism is caused by genetic factors.[108]

Taken together, twin studies have consistently demonstrated that narcissism and the nine other personality disorders are surprisingly more strongly influenced by genetic effects than almost any other type of mental disorder.[109]

---

[108] Gregory W. Lester, *Personality Disorders: Advanced Diagnosis, Treatment & Management* (PESI Publishing, 2018).; Paris, *Fads and Fallacies in Psychiatry* (2023).

[109] Yi lu Luo and Huajian Cai, "The Etiology of Narcissism: A Review of Behavioral Genetic Studies," in *Handbook of Trait Narcissism: Key Advances, Research Methods, and Controversies*, ed. Anthony D. Hermann, Amy B. Brunell, and Joshua D. Foster (Springer International Publishing, 2018), 149-156.; Svenn Torgersen et al., "A Twin Study of Personality Disorders," *Comprehensive Psychiatry* 41, no. 6 (2000): 416-425.; J. Paris, *Treatment of Borderline Personality Disorder: A Guide to Evidence-Based Practice*, 2nd ed. (The Guilford Press, 2020).; Mark Zimmerman, "Overview of Personality Disorders," September 2023, accessed March 31, 2024, https://www.merckmanuals.com/professional/psychiatric-disorders/personality-disorders/overview-of-personality-disorders.

Scientific studies have disproved the popular theories that suggest the events and circumstances of childhood create narcissism, and research has proven that parenting style contributes very little to narcissism[110] and that the relationship between parenting practices and adult narcissism is too small to be considered significant.[111]

The bottom line is this: empirical research has concluded that narcissism is *not* a direct consequence of child abuse or bad parenting[112] and that narcissistic personality disorder cannot develop unless an individual has a genetically predetermined trait profile that makes them susceptible to developing the disorder.[113]

Similar to many genetic disorders, narcissism is polygenic, meaning that it is caused by multiple genes, not just one. In other words, there is no such thing as "the narcissistic gene."

---

[110] Jessica Yakeley, "Current Understanding of Narcissism and Narcissistic Personality Disorder," *BJPsych Advances* 24 (2018): 305-315.

[111] Joshua D. Miller, Thomas A. Widiger, and W. Keith Campbell, "Vulnerable Narcissism: Commentary for the Special Series 'Narcissistic Personality Disorder – New Perspectives on Diagnosis and Treatment'," *Personality Disorders: Theory, Research, and Treatment* 5, no. 4 (2014): 450-451.

[112] Guo Ma et al., "Genetic and Neuroimaging Features of Personality Disorders: State of the Art," *Neuroscience Bulletin* 32, no. 3 (2016): 286-306.; Joel Paris, *Myths of Childhood* (Routledge, 2014).

[113] J. Paris, *A Concise Guide to Personality Disorders* (American Psychological Association, 2015).

Neuroscientific findings have also estimated the heritability of narcissistic personality disorder to be up to 77%.[114] Twin studies have also estimated the heritability of narcissistic personality disorder to be up to 77%.[115] These data findings reveal a high degree of statistical significance.

It is extremely important to note that these same research studies found no *shared* environmental influence. *Non-shared* environmental influence refers to any environmental factors that make twins different from one another in the environmental sense, meaning the way environments influence people differently, regardless of genes. The fact that narcissistic twins who have never shared a common environment have a 77% concordance rate reveals a major genetic influence over narcissism regardless of environment.[116] Genetic research has certainly proven that all personality traits are heritable, just like all physical traits

---

[114] Guo Ma et al., "Genetic and Neuroimaging Features of Personality Disorders," *Neuroscience Bulletin* 32, no. 3 (2016): 286-306.

[115] Ted Reichborn-Kjennerud, "The Genetic Epidemiology of Personality Disorders," *Dialogues in Clinical Neuroscience* 12, no. 1 (2010): 103-114.

[116] Cassandra R. Beam et al., "How Nonshared Environmental Factors Come to Correlate with Heredity," *Developmental Psychopathology* 34, no. 1 (2022): 321-333.; Gwen W. Lester, *Borderline, Narcissistic, Antisocial, and Histrionic Personality Disorders* (PESI Publishing, 2024).; Paris, *Fads and Fallacies in Psychiatry* (2023).

are heritable,[117] and this includes narcissistic personality traits.[118]

What's in a Gene?

A *gene* is a segment of DNA that instructs cells how to function. Genes are made of DNA, and DNA contains the biological instructions that allow for the development, growth, and reproduction of life.

Variation accounts for features in offspring that are different from the parental type. *Genetic inheritance* is the process of genetic information being passed down from parents to their offspring. *Spontaneous mutations* are random alterations or changes that can cause certain genetic conditions and disorders to be expressed even when these conditions and disorders do not run in families. Genetic information inherited from parents influences various aspects of an individual's *phenotype*, which is the set of observable characteristics resulting from the interaction of a genotype with an environment.

---

[117] Siddhartha Mukherjee, *The Gene* (Scribner, 2017).

[118] Nicholas S. Holtzman, "Did Narcissism Evolve?" in *Handbook of Trait Narcissism: Key Advances, Research Methods, and Controversies*, ed. Anthony D. Hermann, Amy B. Brunell, and Joshua D. Foster (Springer International Publishing, 2018), 173-181.

A phenotype includes physical traits, behavior, and susceptibility to certain diseases, disorders, and conditions. However, certain alterations of genes—variants or mutations—are produced in every generation of human beings, so disorders are not always guaranteed to be transmitted from one generation to the next, even though some disorders do run in families. This explains why some family members can have a disorder that historically runs in their family, while other family members do not have that same disorder.[119] There are also different possibilities of inheritance patterns, which account for why some offspring inherit disorders from previous generations and some do not.[120]

Genetic *disorders* occur when you inherit an altered gene from your parents that increases your risk of developing a particular disorder. A genetic disorder can be passed down through a familial line, but again, it's not guaranteed that a child will inherit all the same conditions as their parents.

There are all kinds of genetic conditions and disorders that are not inherited by every single

---

[119] Siddhartha Mukherjee, *The Gene* (Scribner, 2017).
[120] Genetic Alliance, *Understanding Genetics: A New York, Mid-Atlantic Guide for Patients and Health Professionals* (2009) https://pubmed.ncbi.nlm.nih.gov/23304754/.

member of a family. This is often due to varying inheritance patterns and the phenomenon mentioned above known as *spontaneous mutation*, which is, quite simply, a mutation or alteration that spontaneously occurs.[121]

What does all this mean? Quite simply, if someone in your family has a personality disorder like narcissism, it is not a given that you will automatically have a personality disorder too. This is also the case with other disorders, including autism spectrum disorder, ADHD, and schizophrenia, to name a few.

To sum it up, research has shown that just like physical traits, personality traits are undeniably linked to genes.[122]

Genes Are Not Good or Bad

Although it is common to think "bad" or "hopeless" or "doomed" when we hear a behavioral

---

[121] A. Alfano, "Where Does Autism Come from When It Doesn't Run in the Family?" *Labdish*, 2016.

[122] F. L. Coolidge, L. L. Thede, and K. L. Jang, "Heritability of Personality Disorders in Childhood: A Preliminary Investigation," *Journal of Personality Disorders* 15, no. 1 (2001): 33-40.; Yilu Luo and Huajian Cai, "The Etiology of Narcissism," in *Handbook of Trait Narcissism*, ed. Anthony D. Hermann, Amy B. Brunell, and Joshua D. Foster (Springer, 2018), 149-156.; Guo Ma et al., "Genetic and Neuroimaging Features of Personality Disorders," *Neuroscience Bulletin* 32, no. 3 (2016): 286-306.; Siddhartha Mukherjee, *The Gene* (Scribner, 2017).; Svenn Torgersen et al., "A Twin Study of Personality Disorders," *Comprehensive Psychiatry* 41, no. 6 (2000): 416-425.

explanation that includes the word *genetic* as a causal factor, the point of genetic studies, and the point of a discussion about genetics, is really not to focus on something we can't do anything about. It's about seeking greater understanding. Not unsurprisingly, when it comes to traits or abilities we consider "good," such as athletic prowess or a knack for mathematics, genetics becomes viewed as a wonderful thing.

One way to get past the tendency to label genetics as either good or bad is to change our reaction to the term. *Genetic* does not mean something is cast in stone. Genetic is not synonymous with *non-malleable*. In fact, genetic technologies have enabled us to intentionally alter some genes in order to change their function. This does not mean we can burrow into a narcissist's genome and start manipulating the pathological traits, turning them either up or down like a volume knob, but it does mean there is *some* room for improvement in many narcissistic individuals if they are willing to seek help. Some are willing. Most are not.

Why Does *Cause* Matter So Much?

There was a time when schizophrenia was treated with interventions designed to repair unresolved childhood trauma. Now that we know that psychosis is hereditary and there is a strong genetic link, we treat schizophrenia with antipsychotic medication because medication is a better intervention than a corrective emotional experience with a psychodynamic or humanistic therapist. Contrast this with attachment trauma, where a corrective emotional experience is exactly what is needed to repair ruptured attachment. Medication can't correct or ameliorate attachment ruptures, but schizophrenics would be lost without medication.

Take autism as another example—a disorder on the receiving end of intense focus in popular media. Ideas and theories on the cause of autism spectrum disorder have ranged anywhere from the controversial to the downright ridiculous. There was a time when autism experts believed that cold, unemotional parents were the cause. We now know there is a strong genetic link.

When we look at the impact that childhood trauma has on the development of narcissism, the data reaches the same conclusions. For example, just like with schizophrenia and autism, the latest scientific research on narcissism has shown that environmental factors are simply not significant *enough* to produce the development of personality disorders, including narcissistic personality disorder. A predisposed genetic profile *must* exist in order for pathological narcissism to develop.[123]

What matters is not so much that one research or academic field is considered right while another is considered wrong; what matters is that disorders are viewed correctly so that they can be treated correctly. If childhood trauma is not the cause of narcissistic personality disorder, then trying to reduce or cure narcissism by treating it like a childhood trauma disorder will prove unsuccessful. And this has been the case historically. So how do we effectively treat narcissism? In order to answer that question, we need to understand what's going on in a narcissistic brain.

---

[123] J. Paris, *A Concise Guide to Personality Disorders* (2015).

Chapter Eight
The Narcissistic Brain

Normal Brains and Narcissistic Brains

The human brain is essentially a prediction machine. It wants to predict outcomes in order to budget energy and ensure our survival.[124] Neurotypical or *normal* brains, for the most part, make accurate enough predictions about the environment. For example, the sensory data from the outside world—sight, smell, sound, taste, and touch—can be reasonably accurately perceived by your brain. When you're stressed, this perception becomes less accurate. When you're traumatized, this perception becomes even less accurate. When you're disordered, this perception becomes abnormal.

A narcissistic brain is an abnormal brain. It does not perceive things in the same way that normal or overly stressed-out brains perceive things. Two neurotypical individuals experiencing the same sensory data can communicate effectively based on a similar baseline response to their external

---

[124] Lisa Feldman Barrett, *Seven and a Half Lessons about the Brain* (Mariner Books, 2021).

environment and to each other. However, if one of their brains is disordered in the way a narcissistic brain is, interpersonal communication is likely to break down or implode. The colloquial term for this resultant breakdown is *drama*. While this is not a very scientific term, every functioning adult can imagine precisely what this might mean. Narcissistic brains make predictions using pathological escalations of drama rather than collaboration and problem-solving the way normal brains do.[125]

The Narcissistic Brain Is Not a Traumatized Brain

When traumatized or chronically stressed brains make predictions based on threat responses often enough, this can eventually result in the development of unconscious beliefs related to fear and shame. These inaccurate beliefs can motivate some defensive behaviors and tendencies that may require some adjustment; however, this is not the same thing as possessing personality traits such as cruelty, deceit, callousness, arrogance, and entitlement—far from it.

---

[125] Gregory W. Lester, *Personality Disorders: Advanced Diagnosis, Treatment & Management* (PESI Publishing, 2018).

The core beliefs that narcissists possess and the core beliefs that develop after trauma are completely unrelated. And traumatized and stressed-out individuals can change. Upon making new neural connections, the traumatized brain can begin to make more accurate perceptions once again. People who have suffered trauma early in life or later in life and who are not narcissistic, with the right interventions and support, can change their brains and recover.[126]

Narcissists should never be lumped in with those who are stressed out or are trauma victims. Trauma victims and survivors who are not personality disordered possess sufficient flexibility in their personality trait structure. Their beliefs about themselves and others can change, including beliefs that have been influenced by their trauma and have led to shame, self-blame, guilt, feelings of defectiveness, uncertainty in regard to responsibility, and sensations of lack of safety.[127] Narcissistic brains, on the other hand, with the same learning and recovery interventions that are utilized to heal trauma, don't recover from

---

[126] Patricia A. Resick, Candice M. Monson, and Kathleen M. Chard, Cognitive Processing Therapy for PTSD (The Guilford Press, 2017).
[127] Patricia A. Resick, Candice M. Monson, and Kathleen M. Chard, Cognitive Processing Therapy for PTSD (The Guilford Press, 2017).

narcissism. Narcissism does not get resolved through trauma treatment.

What are we supposed to make of this? Are we supposed to continue conceptualizing narcissism through the lens of narcissistic adults being traumatized children who were mistreated in early life or overvalued and spoiled? We can't cling to this narrative while simultaneously acknowledging the reality of the situation: that narcissists don't get better with treatments proven to work with trauma victims and survivors. Quite simply, the bad childhood theory as the *cause* of narcissism doesn't hold up in clinical practice or even in general day-to-day interaction.

I'm still surprised, both professionally and personally, at how so many competent professionals, and the social media world at large, are so willing to accept a theoretical explanation about a disorder that historically has not proved treatable in traditional clinical settings through the use of traditional clinical interventions. Therapists are still at their wit's end when they encounter a narcissist. People in relationships with narcissistic individuals beg for them to own their part of relational problems and plead with them to get help, but narcissists don't own their part and usually will not seek help. On the rare occasion when a narcissist will seek help, the help doesn't help.

Why?

When a narcissist talks about their childhood in therapy, nothing changes. *They* don't change. However, when trauma survivors process their unresolved childhood wounds, they *do* change. There are so many indicators that trauma is not the cause of narcissism, yet some people are absolutely certain that trauma is the reason why narcissists think, feel, and behave the way they do.

The Necessity of Changing Our Perception of Narcissism

There is now enough empirical data to motivate us all to think about narcissism differently. This is a theme that has permeated the study and treatment of mental disorders throughout history. The root cause of personality disorders has historically been attributed to one thing when it turns out to have been something altogether different. As more data comes in, theories are adjusted—that's the evolutionary scientific process in a nutshell.

Faulty Wiring in the Narcissist's Brain

The brain is the organ of our personality. If someone has an unstable personality, it's because of their brain. Narcissistic-personality-disordered individuals are born with deficiencies and impairments in the structure and functioning of their brains. Compared to neurotypical brains, narcissistic-personality-disordered individuals have a decreased number of neurons in certain brain regions that have the potential to develop into mature neurons later in life.[128] The brain contains more neurons at birth than it needs because some will die over time. And some disorders are affected by the way neurons die out or don't increase with development. Narcissism seems to be one of those disorders.

There are identifiable defects in the wiring of narcissistic brains. No doubt about it. Neuro-imaging studies such as CT, MRI, fMRI, EEG, and PET did not exist when narcissism was first being theorized by mental health professionals. Now that we can see the brain in action—literally—and record brain activity, it is undeniable that the narcissistic brain contains multiple structural

---

[128] Gregory W. Lester, *Borderline, Narcissistic, Antisocial, and Histrionic Personality Disorders* (PESI Publishing, 2024).

and functioning anomalies that are not present in normal brains.

Deficits and Impairments in the Narcissist's Brain

Using neuroimaging technologies to measure the structural volume of the insular cortex in individuals with narcissism, the results show that narcissists have a significant reduction of gray matter.[129] This is noteworthy because the anterior insular cortex has been established as the neural region where empathy originates.[130] Neurons exist in this region that are linked to empathy and self-awareness,[131] two traits that narcissists do not possess or are seriously deficient in. The anterior insular cortex is also associated with facial emotion recognition.[132] Studies revealed that

---

[129] Frank R. George and Derek Short, "The Cognitive Neuroscience of Narcissism," *Science Publications* (2017): 1-14.; Lars Shulze et al., "Gray Matter Abnormalities in Patients with Narcissistic Personality Disorder," *Journal of Psychiatric Research* 47, no. 10 (2013): 1363-1369.

[130] Xiaosi Gu et al., "Anterior Insular Cortex Is Necessary for Empathetic Pain Perception," *Brain* 135, no. 9 (August 29, 2012): 2726–35, https://doi.org/10.1093/brain/aws199.

[131] (Henry C. Evrard, Thomas Forro, Nikos K. Logothetis. Von Economo Neurons in the Anterior Insula of the Macaque Monkey. *Neuron*, 2012; 74 (3): 482 DOI: 10.1016/j.neuron.2012.03.003

[132] Frank R. George and Derek Short, "The Cognitive Neuroscience of Narcissism," *Science Publications* (2017): 1-14.; Guo Ma et al., "Genetic and Neuroimaging Features of Personality Disorders," *Neuroscience Bulletin* 32, no. 3 (2016): 286-306.

narcissists who have been given the task of facial emotion recognition performed poorly compared to individuals who are not narcissistic as a result of a narcissist's empathy deficits.[133]

Neuroimaging findings have also revealed that a normal-functioning anterior insular cortex is associated with interceptive awareness, general emotional processing, trust and cooperation, and other qualities that narcissists lack or violate frequently.[134]

Irregularities in the prefrontal regions of the brain, including reduced frontal cortical thickness and volume and weakened frontostriatal connectivity,[135] which is a critical area for higher level processing, judgment and decision-making, social cognition, aggression control, and emotion regulation, were also found in narcissists using neuroimaging studies.[136] There is also decreased cortical

---

[133] Guo Ma et al., "Genetic and Neuroimaging Features of Personality Disorders," *Neuroscience Bulletin* 32, no. 3 (2016): 286-306.

[134] Frank R. George and Derek Short, "The Cognitive Neuroscience of Narcissism," *Science Publications* (2017): 1-14.; S. Park, J. Cho, and Y. Huh, "Role of the Anterior Insular Cortex in Restraint-Stress Induced Fear Behaviors," *Scientific Reports* 12 (2022), https://doi.org/10.1038/s41598-022-10345-2.

[135] S. Ash, D. Greenwood, and J. P. Keenan, "The Neural Correlates of Narcissism: Is There a Connection with Desire for Fame and Celebrity Worship?" *Brain Science* 13 (2023), https://doi.org/10.3390/brainsci13101499.

[136] Frank R. George and Derek Short, "The Cognitive Neuroscience of Narcissism," *Science Publications* (2017): 1-14.; Nenadic, Lorenz, and Gaser, "Narcissistic Personality Traits and Prefrontal Brain Structure."(2021).

volume in narcissistic brains of the left medial prefrontal cortex and right postcentral gyrus, and decreased cortical thickness of the right inferior frontal cortex.[137]

Furthermore, there is decreased cortical thickness and volume in the social brain networks in general in narcissistic personalities, which causes impaired social cognition that affects empathy capacity, emotion regulation, and the ability to control one's behavior.[138] Decreased cortical thickness and volume in the dorsolateral prefrontal cortex limits a person's ability to recognize their own emotions and identify why they are experiencing those emotions, and this decrease is found in the brains of narcissists as well.[139] Reduced cortical thickness in these regions also results in extreme sensitivity to feedback from others, especially negative feedback. This would explain why narcissistic personalities crave admiration and positive attention.[140]

While certain regions of the brain are implicated in narcissism and can be blamed for why some traits are excessively pathological and others

---

[137] Ash, Greenwood, and Keenan, "The Neural Correlates of Narcissism" (2023).
[138] Ash, Greenwood, and Keenan, "The Neural Correlates of Narcissism" (2023).
[139] Ash, Greenwood, and Keenan, "The Neural Correlates of Narcissism" (2023).
[140] Ash, Greenwood, and Keenan, "The Neural Correlates of Narcissism" (2023).

are pathologically diminished or deficient, narcissism is really a whole-brain issue. There is no such thing in real life as the fabled *triune brain*, where we have a separate reptilian, emotional center, and rational cortex part of the brain that regulates the reptilian and emotional parts of the brain. This is a fanciful metaphor for the brain that was never meant to be taken literally.[141]

Quite simply, the brain doesn't work that way. All brain regions function together as a whole. Narcissism is therefore linked to a wide network of brain regions rather than a single brain area, and this widely distributed network works collectively, albeit abnormally, to cultivate and maintain the disorder that we refer to as narcissistic personality disorder.[142]

Narcissists and Their Environment

Are narcissists unaffected by their environment? Of course not.

Genes play a crucial role in building a human being's brain wiring, even before birth.[143] Because of this, there are aspects of personality that are

---

[141] Lisa Feldman Barrett, *Seven and a Half Lessons about the Brain* (Mariner Books, 2021).

[142] Eduardo J. Santana, "The Brain of the Psychopath: A Systematic Review of Structural Neuroimaging Studies," *Psychology & Neuroscience* 9, no. 4 (2016): 420-443.

[143] L. F. Barrett, *Seven and a Half Lessons About the Brain* (Mariner Books, 2021).

built into our genes before we are even born. That being said, we cannot attribute personality to nature alone. Environment also plays a role, but not in the ways that most people have been taught and not in the ways that most people assume. This is where the subject of personality disorders like narcissism gets really interesting, and where the data show surprising results.

Both your genotype and your environment are critically important. Just as there are relationships between genes, there are relationships between genes and the environment.[144] It's crucial to remember that genes don't function to cause disorder on purpose[145]—they just do sometimes. When a disorder like narcissism is present, it affects the way any environment is experienced, interpreted, and dealt with. This has to do with something called *gene-environment interaction*.

Gene-environment interaction refers to the ways in which genetic factors control a person's sensitivity to their environment and therefore influence their response to it.[146] When personality traits are so

---

[144] J. Paris, *Treatment of Borderline Personality Disorder: A Guide to Evidence-Based Practice*, 2nd ed. (The Guilford Press, 2020).

[145] Siddhartha Mukherjee, *The Gene* (Scribner, 2017).

[146] Ted Reichborn-Kjennerud, "The Genetic Epidemiology of Personality Disorders," *Dialogues in Clinical Neuroscience* 12, no. 1 (2010): 103-114.

genetically fixed that they are highly inflexible, that's disorder, and while that disorder is caused by genes, it is certainly intensified by environmental factors.

Remarkably and surprisingly, scientific data has revealed that the environments that worsen narcissistic traits are not adverse childhood environments.[147] Here is what is even more surprising: Longitudinal studies that span at *least* forty years have shown that due to genetic tendencies, narcissists are naturally inclined to *nurture* their own narcissism, not only by manipulating their environments, but by *choosing* environments and relationships—even during childhood—that work to cultivate and maintain their excessive and deficient personality traits.[148] When you examine collective longitudinal studies, you will find that *any* environment for a narcissist seems to lead them down a path of acting out their biologically based temperament and genetically fixed character traits.

Of course, I'm not suggesting that children have the freedom or ability to choose their early environments and relationships, and many children

---

[147] Yilu Luo and Huajian Cai, "The Etiology of Narcissism," in *Handbook of Trait Narcissism*, ed. Anthony D. Hermann, Amy B. Brunell, and Joshua D. Foster (Springer, 2018), 149-156.

[148] Cassandra R. Beam et al., "How Nonshared Environmental Factors Come to Correlate with Heredity," *Developmental Psychopathology* 34, no. 1 (2022): 321-333.

predisposed to develop a disorder like narcissism are exposed to adverse environmental experiences early in life. But narcissism is in fact predisposed before any childhood adversity occurs and even if it never occurs. Paradoxically, *every* possible environment becomes a risk factor for narcissism when the genetic factors are present, including environments that are not freely chosen.[149]

Narcissistic personalities experience and contribute to their social environments in different ways based on their genetic predisposition. We have seen this difference occur among biological siblings raised in the same household. If one sibling inherits the genetic predisposition that is based on a certain inheritance pattern or the randomness of spontaneous gene alterations to be narcissistic, and the other sibling does not inherit that same predisposition, the one who does inherit it will select and evoke maladaptive experiences in their social environment, while the non-narcissistic sibling will select and evoke adaptive environmental experiences.[150]

---

[149] Ted Reichborn-Kjennerud, "The Genetic Epidemiology of Personality Disorders," *Dialogues in Clinical Neuroscience* 12, no. 1 (2010): 103-114.
[150] Cassandra R. Beam et al., "How Nonshared Environmental Factors Come to Correlate with Heredity," *Developmental Psychopathology* 34, no. 1 (2022): 321-333.

Based on this research, it is most accurate to say that genes and environment need to be taken together to complete the whole picture of personality pathology. In a way, you need both. Genetic factors require an environment for trait expression, and the environment requires genetic factors for the varying types of trait expression.[151]

Regardless of the type of environmental factor, narcissism is present from the beginning. And, interestingly, the environment that does *not* seem to be a risk factor for narcissism at all, according to neuroscientific, neurobiological, and behavioral genetics research, is that of being raised in a "certain way" and growing up in a particular kind of family environment.[152]

As stated before, the influence of genes goes beyond genetic inheritance. Genetic inheritance is not linear and simple. It involves variations, multiple genes, randomness, and environments that

---

[151] Robert M. Sapolsky, *Behave: The Biology of Humans at Our Best and Worst* (Penguin Books, 2017).
[152] Cassandra R. Beam et al., "How Nonshared Environmental Factors Come to Correlate with Heredity," Developmental Psychopathology 34, no. 1 (2022): 321-333.; Yilu Luo and Huajian Cai, "The Etiology of Narcissism," in *Handbook of Trait Narcissism*, ed. Anthony D. Hermann, Amy B. Brunell, and Joshua D. Foster (Springer, 2018), 149-156.; Paris, *Myths of Trauma* (2023).

allow for gene expression.[153] All of these factors work together to make trait expression more or less likely. Someone with a predisposition to narcissism may only express trait narcissism because the "environment" within their cells and certain gene variations makes it possible. This predisposition has been shown to occur before someone is even born and sets a person up to be a certain way, but it also requires an external environment for trait expression. However, empirical research has demonstrated that when it comes to the expression of pathological narcissism within a given environment, the environment in and of itself does not so much create the narcissism; rather, the environment that most caters to the traits of narcissism is in fact chosen and sought out by the narcissist based on their genetically fixed traits. When a given environment cannot be chosen, as is the case with minor children, the perception of their environment and how they deal with it are interpreted through their genetically predisposed narcissistic lens. So while environment plays a role in narcissistic trait expression, no environment or environmental experience is dominant or

---

[153] Siddhartha Mukherjee, *The Gene* (Scribner, 2017).

influential enough to actually be considered the *cause* of pathological narcissism.

The Five-Factor Model of Personality

If personality disorders are to be seen as genetic variations of trait deficiencies and trait excesses, it is important to determine what a "normal" personality trait is. The five-factor model of personality is one of the most popular and validated models for assessing personality traits. The personality traits included in the five-factor model are as follows:[154]

1. Neuroticism, which includes anxiety, angry hostility, depression, self-consciousness, impulsiveness, vulnerability.
2. Extraversion, which includes warmth, gregariousness, assertiveness, activity, excitement seeking, positive emotions.
3. Openness to experience, which includes fantasy, aesthetics, feelings, actions, ideas, values.

---

[154] Marijn A. Distel et al., "The Five-Factor Model of Personality and Borderline Personality Disorder: A Genetic Analysis of Comorbidity," *Biological Psychiatry* 66, no. 11 (2009): 1131-1138.; Brendan Weiss and Joshua D. Miller, "Distinguishing between Grandiose Narcissism, Vulnerable Narcissism, and Narcissistic Personality Disorder," in *Handbook of Trait Narcissism*, ed. Anthony D. Hermann, Amy B. Brunell, and Joshua D. Foster (Springer, 2018), 3-13.

4. Agreeableness, which includes trust, straightforwardness, altruism, compliance, modesty, tendermindedness.
5. Conscientiousness, which includes competence, order, dutifulness, achievement striving, self-discipline, deliberation.

The five-factor model has been used in clinical research to demonstrate that disordered personalities, that is, personalities with abnormal trait deficiencies and trait excesses like narcissistic personalities, think, feel, and behave the way they do as a result of the presence of genetically affected, extreme forms of normal personality traits like those assessed in the five-factor model of personality.[155] (See table 3.)

For example, narcissists are very low in two traits within the five-factor model (See table 3): conscientiousness and agreeableness.[156] These lower trait levels in individuals are accounted for not by bad childhoods, but by genetic variance.[157] This means that genetic variations in trait levels that are inherited and heritable (two different

---

[155] Marijn A. Distel et al., "The Five-Factor Model of Personality and Borderline Personality Disorder," *Biological Psychiatry* 66, no. 11 (2009): 1131-1138.
[156] Paris, *Myths of Trauma* (2023).
[157] Paris, *Fads and Fallacies in Psychiatry* (2023).

things) determine the difference between normal personality traits and pathological personality traits, rather than childhood adversity.[158]

So because empirical data proves that genes contribute significantly to the variance in personality traits, both in terms of normal and pathological trait levels,[159] it is inaccurate to attribute the *cause* of personality disorders like narcissism to adverse childhood experiences. We have to take into account the hard data that shows how much genes influence narcissistic traits.[160]

Similarly, because studies have shown that there is not a single parenting behavior or specific childhood experience that has ever predicted the development of narcissism,[161] we have to accept that the data gathered from anecdotes of childhood and theoretical interpretation, in other words, data that is subjective rather than objective, cannot be relied upon any longer to understand the cause of narcissism.

---

[158] Yilu Luo and Huajian Cai, "The Etiology of Narcissism," in *Handbook of Trait Narcissism*, ed. Anthony D. Hermann, Amy B. Brunell, and Joshua D. Foster (Springer, 2018), 149-156.; Paris, *Myths of Trauma* (2023).

[159] Marijn A. Distel et al., "The Five-Factor Model of Personality and Borderline Personality Disorder," *Biological Psychiatry* 66, no. 11 (2009): 1131-1138.

[160] Yilu Luo and Huajian Cai, "The Etiology of Narcissism," in *Handbook of Trait Narcissism*, ed. Anthony D. Hermann, Amy B. Brunell, and Joshua D. Foster (Springer, 2018), 149-156.

[161] Gregory W. Lester, *Personality Disorders: Advanced Diagnosis, Treatment & Management* (PESI Publishing, 2018).

Table 3. Five-factor model of personality and narcissistic personality disorder: trait excesses and deficiencies

FIVE-FACTOR MODEL	EXCESSIVE	DEFICIENT
Neuroticism		
Anxiety		
Angry hostility	X	
Depression		
Self-conscious		X
Impulsiveness	X	
Vulnerability		X
Extraversion		
Warmth		
Gregariousness		
Assertiveness		X
Activity		
Excitement seeking		X
Positive emotions		X
Openness to experience		
Fantasy	X	
Aesthetics		
Feelings		
Actions		
Ideas		
Values		

Agreeableness		
Trust		X
Straightforwardness		X
Altruism		X
Compliance		X
Modesty		X
Tender-mindedness		X
Conscientiousness		
Competence		
Order		
Dutifulness		X
Achievement striving		
Self-discipline		X
Deliberation		X

There is good news. Because we now know beyond a shadow of a doubt that narcissism is a genetic, neurological disorder and not a trauma-related condition resulting from stunted psychological development in childhood that extends into adulthood,[162] we can treat it as such. And when we treat narcissism like the genetic, neurological

---

[162] Gregory W. Lester, *Personality Disorders: Advanced Diagnosis, Treatment & Management* (PESI Publishing, 2018).

disorder that it is, treatment improves by about 100 percent.[163]

Revisiting Narcissism and Comorbidity

As mentioned before, *comorbidity* refers to two or more conditions or disorders existing simultaneously within an individual that are completely independent of each other. Many narcissists also have comorbid depression and substance-use disorders and can have trauma-related disorders like PTSD in addition to narcissistic personality disorder, but these disorders are *not* related to narcissism, nor do they contribute to the *cause* of narcissism.

Gaining a clear understanding of how comorbidity works will hopefully clarify how a narcissist can come from a traumatic background and meet the criteria for PTSD or depression but also be narcissistic, and that the narcissism is neither caused by nor affected by comorbid PTSD or depression.

Comorbid conditions such as PTSD and depression *can* be effectively treated in an individual who also has narcissistic personality disorder while the narcissism remains unchanged. Results

---

[163] Gregory W. Lester, *Borderline, Narcissistic, Antisocial, and Histrionic Personality Disorders* (PESI Publishing, 2024).

from a replicated clinical trial that examined the effects of prolonged exposure and cognitive processing therapy[164] for individuals with personality disorders revealed that the use of cognitive processing therapy and prolonged exposure created improvements in personality-disordered individuals who have comorbid PTSD and depressive symptoms, but the therapy did *not* improve the disordered features of their personality.[165]

In short, trauma interventions in personality-disordered individuals have been effective in alleviating symptoms associated with comorbid symptomatic conditions like depression and post-traumatic stress disorder in these individuals. Just as a narcissist who also happens to have depression can benefit from taking an antidepressant medication resulting in dissipated depressive symptoms, a narcissist who also has comorbid post-traumatic stress disorder can participate in treatment for PTSD and experience an alleviation or reduction of PTSD symptoms. But what's important to note is this: under the circumstances above, regardless of the evidence-based

---

[164] Patricia A. Resick, Candice M. Monson, and Kathleen M. Chard, Cognitive Processing Therapy for PTSD (The Guilford Press, 2017).

[165] Patricia A. Resick, Candice M. Monson, and Kathleen M. Chard, Cognitive Processing Therapy for PTSD (The Guilford Press, 2017).

interventions that successfully resolve trauma in individuals who also meet the criteria for narcissistic personality disorder, the narcissistic traits remain unchanged, fully intact, and as pervasive and enduring as ever.

Let me repeat this: a narcissist with comorbid PTSD or depression can receive interventions that successfully treat PTSD and depression and experience a resolution or reduction of symptoms of PTSD or depression while their narcissism remains fully intact, unchanged, inflexible, enduring, and as pervasive as ever. This is more evidence that narcissism is not caused by trauma, depression, or any other disorder that a narcissist may also have in addition to narcissism. We know narcissism is the disorder in question when the presentation has been going on since early childhood and is pervasive (meaning across the board in all or most areas of life) and is persistent (meaning the rule rather than the exception). Comorbidity is interesting, but it can distract from the root of the narcissistic problem.

Chapter Nine
Early Childhood Trauma and Narcissism

The human attachment system is a neuro-biological system that requires parental attunement, comfort, enjoyment, engagement, guidance, and love in order to develop properly in normal brains. Failure to experience these emotional necessities can result in the development of trauma-related symptoms and conditions that can persist throughout life and impact our most significant relationships as adults.[166] It is a scientific fact that when social-emotional needs are not adequately met, babies develop intellectual difficulties as well as difficulties related to concentration and self-control.[167] Is there an exception to this? Yes: those individuals who never developed the attachment wiring to begin with.

Attachment is more about *emotional* safety than anything else. If a child with a normal-functioning brain is neglected or experiences compromised safety, that child's autonomic nervous system

---

[166] Knipe, J. *EMDR Toolbox: Theory and Treatment of Complex PTSD and Dissociation.* New York, NY: Springer Publications (2018).
[167] Lisa Feldman Barrett, *Seven and a Half Lessons about the Brain* (Mariner Books, 2021).

responds to close relationships in the same way it would if they were in imminent physical danger. Recent research has shown, for example, that a parent who is consistently verbally aggressive toward a child with normal neurobiology increases the likelihood of that child developing symptoms of anxiety, depression, hypervigilance, concentration difficulties, and irritability in adulthood even *more so* than if that parent were to have been physically abusive.[168] But as previously discussed, narcissists don't exactly have "normal" neurobiology.

How Attachment Goes Wrong in Non-Disordered Individuals

Parent-child attachment ruptures occur for a number of reasons. Parents who suffer from mental disorders, who are preoccupied with their own unresolved traumatic pasts, who didn't receive what they needed from their own parents in childhood, who struggle with addiction or substance abuse, and who don't want to be parents, who didn't plan to be parents, or couldn't be good parents due to socio-economic or cultural circumstances can seriously,

---

[168] Lisa Feldman Barrett, *Seven and a Half Lessons about the Brain* (Mariner Books, 2021).; Knipe, J. *EMDR Toolbox: Theory and Treatment of Complex PTSD and Dissociation.* New York, NY: Springer Publications (2018).

even if unintentionally, rupture attachment bonds with the children who depend on them. However, this doesn't pave the way for narcissism to develop. Narcissistic children are different from the start. These examples of attachment adversity don't imprint onto them the way they do in children with normal neurobiology.[169]

Our mental and physical well-being, for the most part, depends on our ability to recognize patterns and anticipate outcomes.[170] That's how the brain works; it's basically a survival and prediction machine. Typically, we have control over the memories we recall, and we can control when they start and when they stop. But trauma hijacks this ability to the point that we are no longer in control of what we experience because we still perceive a threat long after the threat is gone. When this happens, we will try to avoid or push certain memories away, and this only makes matters worse over time. Prolonged stress produces prolonged mental and physical consequences and makes it seem like threats that have already taken place are still happening in the present.[171]

---

[169] J. Paris, *A Concise Guide to Personality Disorders* (2015).
[170] M.J. Gerson, *Child Abuse and Trauma*, Westlake Village, CA: Institute of Advanced Psychological Studies (2021).
[171] Shapiro, F. *Eye Movement Desensitization and Reprocessing (EMDR) Therapy: Basic Principles, Protocols, & Procedures*. 3rd ed. New York, NY: The Guilford Press (2018).

In terms of narcissists, however, all of the above does not really apply. They are not hijacked by fear and shame as result of perceiving threat to their relational safety. A narcissist can very well suffer from PTSD if they witness a traumatic event or they are exposed to their own threat of death, for example, but this is not the same thing as relational trauma or attachment trauma. Based on the deficits and impairments in the brains of narcissists discussed in the previous chapter, it is not really possible for them to attach in the same ways that neurotypical brains attach to one another. Therefore, it is not really even possible for them to suffer attachment trauma.

Attachment History Does Not a Narcissist Make

Attachment relationships leave imprints on our brain and nervous system. Adults often find themselves in situations in adulthood that eerily resemble their childhood attachments. For example, a person might marry someone who embarks on an extramarital affair ten years into the marriage only to later discover their mother began an extramarital affair a decade into her own marriage. This is completely different from personality-disordered behavior because it is not conscious or deliberate behavior. It's an unconscious process.

Narcissists, on the other hand, behave the way they behave consciously and on purpose. Their exploitation and manipulation can seem like it's automatic and habitual due to the frequency of it,[172] but they do consciously and deliberately exploit and manipulate others even when they are unprovoked and have no need to defend themselves.

In normal individuals, intergenerational transmission of specific attachment repetitions and behaviors is unconscious and often unpreventable. Normal people attach, and even if they insecurely attach, they seek out the attachment pattern experienced as children with parents or caregivers in adult relationships because they are committed to that familiar pattern on an unconscious level, whether they want to be or not. Pathological narcissists do not seek out situations to work through old baggage. They are disinterested in

---

[172] George K. Simon Jr., *Character Disturbance: The Phenomenon of Our Age* (Parkhurst Brothers Publishers, 2011).

their relationships from childhood to begin with due to their neurobiological abnormalities and impairments.[173] Narcissists don't avoid attachment, nor does their attachment fail; they simply fail to attach. They are the equivalent of tape without adhesive. They don't care, they don't share, they don't love.[174] For a narcissist, there's nothing to work out or through later on because there was nothing of value or meaning to them about their earlier relationships to begin with.

It has been theorized that narcissists have an avoidant attachment style. Avoidant attachment is typically the result of suppressed emotions and intimacy intolerance motivated by deep-seated fears of rejection. Narcissists can't have an avoidant attachment style due to fear of intimacy or rejection because they don't attach to begin with. It is more accurate to say that narcissists have a *disinterested* detachment style. They would need to be taught how to attach because they don't attach naturally like most people.

---

[173] Edward Bleiberg, *Treating Personality Disorders in Children and Adolescents: A Relational Approach* (The Guilford Press, 2001); Paulina F. Kernberg, Alan S. Weiner, and Karen K. Bardenstein, *Personality Disorders in Children and Adolescents* (Basic Books, 2000).

[174] Donald J. Robinson, *Disordered Personalities*, 3rd ed. (Rapid Psychler Press, 2005).

Narcissism, Trauma, and Attachment

Trauma is not a mental disorder; it is an adaptive response to environmental stimuli.[175] Let me repeat that: trauma is *not* a disorder. Narcissism is a mental disorder. Traumatic stress responses are adaptive responses to emotional distress. It is extremely important for a trauma victim or trauma survivor's overall well-being to understand that being traumatized is an adaptive response that comes with the territory when suffering ruptures by attachment figures. It is also important to understand that attachment trauma, and trauma in general, has absolutely nothing to do with the development of narcissism or any other personality disorder.[176]

Someone being raised by a parent or caregiver with narcissistic personality disorder, for example, would not be able to avoid being traumatized by the attachment bond. Attachment trauma is inevitable. However, while being raised by someone who fits the bill for a personality disorder like narcissism can and most likely will traumatize you and impact your attachment style, it will *not* turn you into a narcissist like them.

---

[175] Lisa Feldman Barrett, *Seven and a Half Lessons about the Brain* (Mariner Books, 2021).
[176] Joel Paris, *Myths of Trauma* (2023).

Unlike trauma, a mental disorder like pathological narcissism relates more to an excess or a deficiency in *overall* functioning. It pertains to characteristics rather than symptoms, and it doesn't run a course; rather, it's always there. It's pervasive, meaning that it impacts most or even all aspects of life rather than just one area or one system. When attachment trauma and narcissism are compared side by side, they really have nothing in common. (See table 4.)

Table 4. The Differences between Attachment Trauma and Narcissistic Personality Disorder

Attachment Trauma	NPD
Condition of stress & trauma.	Disorder of the person.
Symptomatic behavioral condition.	Enduring pattern.
Erroneous thoughts & beliefs related specifically to traumatic relationships.	Erroneous thoughts & beliefs related to reality in general.
Self-blaming that leads to guilt and shame.	Other blaming that is void of guilt and shame and that leads to justification and rationalization.
Flexibility in personality works to change beliefs about trauma and allows for the development of a secure attachment style.	Inflexibility in personality works to keep beliefs fixed. There is no development of secure attachment because there is no attachment to begin with.
Trauma triggers are activated when relational danger is perceived.	NPD individuals are triggered when they aren't getting what they want or aren't being sufficiently admired. They do not perceive relationships as something to fear.
Trauma responses force the brain to protect from threats it believes are present.	NPD forces the brain to edit or block out anything that is not consistent with an NPD individual's desired worldview.
Post-traumatic beliefs don't fit well with one's present environment because the beliefs are related to danger that is no longer present.	NPD beliefs don't fit well with one's past or present environment because they are fixed and inflexible regardless of past experience.
Attachment trauma activates fear and anxiety.	NPD activates rage and impulsivity.

Attachment trauma is a nervous system condition that involves chronic stress in the nervous system due to the perception of threat.	NPD is a trait deficiency disorder that involves the expression of excessive and deficient levels of pathological personality traits, including arrogance, entitlement, inequality, callousness, grandiosity, manipulation, deception, antagonism, uncooperativeness, hostility, and exploitation.
Attachment trauma can be successfully resolved with science-based trauma interventions. Normal baseline of functioning can be restored or be experienced for the first time.	NPD features and traits are not alleviated with science-based trauma interventions. Normal baseline of functioning cannot be restored because normal baseline of functioning in NPD is disordered.
Attachment trauma shapes symptoms and perceptions of self, others, and the world related to safety, trust, and esteem.	NPD shapes personality traits and perceptions of self, others, and the world related to grandiosity, the need for admiration, and a lack of empathy.
Attachment trauma activates defenses for the purpose of coping with stress and returning to a level of safety.	NPD activates defense for the purpose of manipulating and returning to a level of control or advantage.
Attachment trauma, like other forms of trauma, is a condition of "non-recovery," meaning something had to happen that has not been completely recovered from.	NPD is a disorder of personality trait deficiency and excess, meaning it was there before anything happened and it is there when nothing happened.

© 2024 Peter Salerno

Defense Mechanisms

Narcissists differ from trauma victims and survivors in how they utilize defense mechanisms. Every human being utilizes defense mechanisms, but the *motivation* for the defenses is an important distinction here. A person suffering from attachment trauma will utilize defense mechanisms such as denial, avoidance, justification, people pleasing, resistance, suppression, or projection, for example, because they are truly protecting themselves from what they perceive as imminent danger within the context of a significant relationship. In this case, their defense mechanisms are nervous system responses that are really just attempts to ensure safety and survival. Relationships can be very threatening for someone who has suffered attachment trauma, and this naturally activates the autonomic nervous system, which causes defensiveness.

On the other hand, when a narcissistic personality implements the same or similar defenses, their motivation, believe it or not, is to sustain existing problems and make new ones—on purpose.[177] The narcissist is not interested in collaboration, which can seem paradoxical because they may act

---

[177] Gregory W. Lester, *Borderline, Narcissistic, Antisocial, and Histrionic Personality Disorders* (PESI Publishing, 2024).

as if they are truly invested in a relationship while their primary motive is actually to create conflict. If confronted with their true motive, they will vehemently deny it. As we have learned, narcissists are inflexible, are seriously lacking in their ability and their willingness to problem solve in relationships, have problematic deficiencies and excesses in their temperament and personality traits, intentionally misrepresent the truth by lying or distorting reality, and are highly inaccurate in their perception of the cause of problems and conflicts.[178]

Anxious Attachment

Anxiously attached individuals continue pursuing affection even after they receive it because they doubt the reliability and authenticity of the affection. Their partners tend to describe them as "clingy," and the anxiously attached often desire consistent reassurance and validation of feelings. They have a tendency to want to be close to their partner in physical proximity as often as possible, which can cause their partner to accuse them of losing sight of their independence

---

[178] Gregory W. Lester, *Personality Disorders: Advanced Diagnosis, Treatment & Management* (PESI Publishing, 2018).

and individuality. They are often preoccupied with fears of being left by their partner or being on the losing end of a break-up because the closeness they desire seems to be misunderstood and pathologized by many of their partners.

Narcissists, on the other hand, do not care about shoring up the affection they receive from others because they don't have the brain wiring to attach in the first place. When they present as anxious or needy in relationships, it is not because they are demanding affection or even seeking it; it is because they are expecting admiration and confirmation. As discussed previously, this has been clearly revealed in neuroimaging findings.

Avoidant Attachment

Avoidantly attached individuals tend to withhold and avoid affection, have difficulty expressing emotions, and typically will not seek help or ask for it unless prompted by someone else. They still have a deep desire to be intimate, so if they are ignored or rejected, they might temporarily become more pursuant of affection until their partner reciprocates; then they will withdraw again. This behavior is the result of fear of intimacy due to how much it can hurt to feel rejected by someone you deeply care for and love.

Avoidantly attached individuals tend to minimize the importance of relationships and their reliance upon others in their lives as a defense. They are often more physically or sexually motivated than emotionally motivated to connect with another person because a sexual/physical connection is much less threatening than emotional intimacy. Deep down, however, the avoidantly attached individual really wants to love and be loved. The vulnerability that is required to love and be loved is what causes the avoidant individual to recoil and avoid the very intimacy they desire.

Narcissists are likewise avoidant when it comes to emotional intimacy, but not for the same reasons as the avoidantly attached individual. Narcissists are not vulnerable, nor are they attachment wounded; they can't attach in the first place. Their avoidance of genuine intimacy is due to their faulty wiring, not their hidden fears. They also do not desire to love and be loved. As stated before, they desire to be admired and expect to be regarded as superior. This is what makes them happy. Love and intimacy do not appeal to the narcissist.

Disorganized Attachment

Individuals with a disorganized attachment style display characteristics of both anxious and avoidantly attached individuals. They typically have very low self-esteem and tend to complicate intimacy because they fear rejection. They inadvertently sabotage relationships by asking for too much and not giving enough. Their intentions are not malicious; they are simply panicked by intimacy because they are terrified of it.

Because disorganized attachment is a combination of anxious and avoidant attachment, narcissists do not fit the bill for this attachment style either, but someone who had a parent who was a narcissist could very well develop a disorganized attachment as a result.

None of the attachment styles above are related to narcissism. To have suffered from attachment trauma, the requirement is that you possess the capacity to attach in the first place. Based on the neurobiological deficits and impairments in the narcissistic brain, it's not possible for a narcissist to have an attachment style, nor is it possible for a narcissist to experience traumatic attachment in the way that normal people do. The traits and tendencies that make up pathological narcissism have nothing to do with attachment ruptures. They have to do with the way the brain is wired from the beginning. For the narcissist, there are no attachment wounds or ruptures to heal or process and recover from because there was no attachment in the first place. As unfortunate and sad as it may seem, narcissists don't need other people for the purposes of attachment. They need people to gratify their own selfish desires. There is no reciprocal relationship with a narcissist because there is no attachment. This is one the most difficult facts to wrap your mind around when it comes to viewing a narcissist objectively—for what and who they truly are—but to understand those with Narcissistic Personality Disorder, it is critically important to do so.

Chapter Ten
Narcissistic Abuse

Attaching to a Narcissist Is Traumatic

My mother and I were the only two people present when my grandmother passed away. She died in the same hospital I was born in; she took her last breath in the same hospital where I took my first breath. I held her hand for an hour after she died. My attachment to her, due to the way she was, was complex to say the least. I was flooded with mixed emotions when she died. She was the matriarch of my family—a dysfunctional matriarch, but a matriarch, nonetheless.

I remember my mom saying: "I was a daughter, and now I'm not. Now I don't have a mother." My heart broke for my mom. But not because of what she said. My heart broke because the sad reality is that my mom *never* had a mother. Not in the true sense of what the word means.

Part of the reason I included my grandmother's story in this book is because I want to encourage people who are or have been in a similar situation to take *less* responsibility for the difficult people in their lives. Sometimes the only solution for a better relationship with certain people is to step away and wait until they decide to do something about themselves—even if they never do. Every other option requires the person who is not to blame and who is not responsible to become overburdened, overtaxed, and emotionally drained. And most of the time, their suffering does not change the narcissistic person or the outcome; thinking this is possible is really just wishful thinking. The reality is that even though there is a genetic predisposition to narcissism, which is no one's fault, these individuals need to behave and treat others better than they do. It is up to *them* to be different, whether they are capable of it or not.

Narcissistic Abuse

Narcissistic abuse has become a catchphrase in popular media. I believe it should be. Although the term is often misused and misunderstood, and it tends to be freely applied as if it identifies some newly discovered or recent type of abuse, the reality is that this form of abuse is nothing new, and in some ways, narcissistic abuse has created a global mental health crisis. The reality is that narcissistic people tend to abuse and manipulate the people closest to them in specific ways, and the trauma that results from being in an abusive relationship with a narcissist is a unique and exclusive form of trauma that goes beyond attachment ruptures and typical PTSD symptoms.[179]

How the manipulation presents can vary from narcissist to narcissist, but the desired result is always the same: to achieve control or advantage for some typically distorted purpose without any regard for the feelings and needs of others. Being on the receiving end of a narcissist's callous behavior on a regular basis constitutes abuse because of what such abuse does to the brain and nervous system when it is so frequent and consistent. If you are frequently and consistently exposed to verbal, emotional, and psychological manipulation tactics that amp up and shut down your nervous system and cause you to feel unsafe, this is equivalent to a million paper cuts. A few paper cuts won't kill you, but enough paper cuts can cause so many micro-traumas that those paper cuts can become extremely hazardous to your health.

Due to the severe degree of conscious and deliberate truth manipulation, deception, devalu-

---

[179] S. L. Brown and J. R. Young, *Women Who Love Psychopaths: Inside the Relationships of Inevitable Harm with Psychopaths, Sociopaths, and Narcissists*, 3rd ed. (Mask Publishing, 2018).

ation, coercion, subterfuge, and countless other tactics narcissists are known for using to achieve their desired results regardless of the consequences to others, victims of narcissistic abuse develop unique trauma symptoms that cause alterations in the structure and functioning of their brains. These changes don't necessarily damage the brain permanently, but they work to immobilize a person's autonomic nervous system and core belief system in ways that are exclusive to abuse suffered at the hands of a pathological narcissist. Therefore, healing from this insidious form of maltreatment requires distinctive interventions.

What Is Traumatic Cognitive Dissonance?

There is strong empirical evidence that individuals who develop trauma-related conditions during and following a relationship with a pathological narcissist experience significant disruptions in pre-existing beliefs.[180] These disrupted beliefs are manifested in inaccurate self-statements that interrupt or pause the normal recovery process. These disrupted beliefs are typically shame-based,

---

[180] Patricia A. Resick, Candice M. Monson, and Kathleen M. Chard, Cognitive Processing Therapy for PTSD (The Guilford Press, 2017).

and they also relate to lack of safety, lack of control, and assuming responsibility for what is not one's fault.

Due to the nature, structure, and functioning of narcissistic personalities, being on the receiving end of their abuse can be brutal and unrelenting. When truth manipulation, including pathological lying, gaslighting, subterfuge, and other forms of deceit and coercion are experienced on a chronic basis, the result is a devastating, disorienting, and bewildering form of complex trauma with a very unique indicator that goes far beyond typical trauma and stress-related symptoms. I call this unique symptom indicator *traumatic cognitive dissonance.*

Traumatic cognitive dissonance can be defined as a distinct form of dissonance that is a direct result of chronic truth manipulation within the context of an intimate relationship with a cluster B personality, like a narcissist, causing your brain to be inundated with simultaneous contradictory beliefs about yourself, your partner, and the true nature, quality, and reality of your relationship. While this is occurring, many victims tend to blame *themselves* for thinking and feeling this way and punish themselves with harsh criticism for not being able to decipher the truth or not being able to just "get over it."

Indicators of Traumatic Cognitive Dissonance:

- Rumination or "racing thoughts" about the relationship make you feel "amped up" and "shut down" at the same time.
- Loss of cognitive control of your behavior and possibly acting out of character as a result (diminished executive functioning).
- Discursive thoughts, feelings, and beliefs related to self-identity (as if you don't know who you are anymore).
- Emotional paralysis/chronic "freeze" response.
- Heightened defensiveness and sensitivity/inability to detect a real threat of safety from a perceived threat of safety, and fearing both.
- Complete loss of self-worth.
- Excessive self-blame.
- Dissociative experiences like staring into space or tuning out/losing time.
- Random reassembling of memories of your abusive relationship which confuse you and cause you to question how you truly feel about your abuser. This can also cause you to question whether you

are even capable of accurately perceiving the relationship dynamics or if you have made the right decision about the status of the relationship (staying or leaving).

- Unwarranted feelings of extreme guilt and shame.
- The feeling or fear that you have "gone crazy" and cannot trust yourself, your perception, or your own judgment.
- A profound sense of emotional isolation.
- Desire for physical isolation while simultaneously feeling a desperate urge to be supported and understood in good company.
- Inability to make even the most basic, trivial everyday decisions.
- Pushing away your instincts and intuition.
- Experiencing constant fear and threat states without an identifiable cause.
- Thought immobilization. Feeling stuck in the same thought without being able to stay focused enough to make a decision or come to any conclusions.
- Difficulty processing communication with others or engaging in dialogue.
- Insomnia or hypersomnia. Or both.

- Psychomotor immobilization (difficulty physically functioning and getting your body to work and move at will).
- Feeling like a fraud or a failure.
- Feeling like an imposter.
- Terrifying vivid nightmares or vivid images throughout the day that come seemingly out of nowhere.
- Worst-case scenario thoughts.
- Inability to disengage from your former partner (looking up their social media accounts, going over past conversations and arguments, arguing "with them" to yourself in real time to try to prove points that were dismissed and defend yourself even after no contact).
- Generalized anxiety.
- Hypervigilance.

Traumatic Cognitive Dissonance and the Stress Response System

In addition to the relentless dissonant thoughts, feelings, and beliefs that constantly intrude upon a victim of narcissistic abuse at random, traumatic cognitive dissonance can also simultaneously activate and maintain a chronic "freeze" response in your autonomic nervous system.

After enduring chronic manipulation and abuse from a narcissistic personality, the brain of a victim assembles multiple neural pathways of fear, shame, and doubt that activate reminders of the abuse and maintain cognitive dissonance.

Here's how it works: the dorsal vagal complex or "freeze" response system is the oldest threat response system in the body.[181] While the dorsal vagal state is designed to promote survival, prolonged activation of this "freeze" response system resulting from chronic fear can lead to the mental haziness and physiological stress that are characteristic of traumatic cognitive dissonance. Dorsal vagal "shutdown" essentially puts the mind and body into a paralysis of sorts. Not only is this chronic immobilization physically and emotionally exhausting, but it's also disorienting and confusing for the victim. It's as if you can't figure out what's wrong and you can't figure out why you can't figure it out.

Traumatic cognitive dissonance is by far one of the most debilitating trauma symptoms after

---

[181] J. E. Gentry, *Forward-Facing Freedom: Healing the Past, Transforming the Present, a Future on Purpose*. Parker, CO: Outskirts Press (2021).

being involved with a personality-disordered partner like a narcissist.[182] Sadly, it is also almost completely undetected, overlooked, misunderstood, or surprisingly, even denied or reframed by the vast majority of mental health professionals who specialize in treating complex trauma.

The Difficulty in Finding Competent Professional Support

If you have tried to seek help or support in the past from a professional, or multiple professionals, to try to make sense of what you've been through with a narcissist, and they were not aware of the distinct form of trauma that I just described and that is unique to being in a relationship with a pathological narcissist, it's very likely that the therapy did not help, and you found yourself once again scouring for answers and desperately seeking support that could finally help you make sense of what you're going through and what happened to you. Maybe you had to become an expert yourself by delving into your own investigative research.

---

[182] S. L. Brown and J. R. Young, *Women Who Love Psychopaths: Inside the Relationships of Inevitable Harm with Psychopaths, Sociopaths, and Narcissists* (2018).

Why Previous Attempts to Get Help Were Not Very Helpful

One of the main points of this book is to bring awareness to the fact that many mental health professionals, some knowingly, some unknowingly, hold theory biases that can prevent effective and competent care. As a victim of narcissistic abuse, it is sadly all too common to end up in therapy with a well-meaning professional who cannot adequately educate survivors on narcissistic abuse or the true nature of the relationship they survived or are still suffering through. Because of this, many mental health professionals cannot intervene in ways that produce lasting recovery and healing in victims and survivors. In this way, the field of mental health has let victims and survivors of pathological relationship abuse down in a major way. The reality is that the vast majority of therapists *cannot* sufficiently validate survivors of narcissistic abuse or effectively treat their distinct trauma symptoms because they weren't taught what they need to know.

The biases and erroneous beliefs that most mental health professionals continue to hold about narcissism inadvertently cause harm to individuals who are victimized by narcissists. If a therapist believes that all human beings are basically the same on the inside, and they assume that even pathological personalities like narcissists are very treatable, can be reasoned with, and for the most part are genuinely willing to collaborate and problem solve, there is a high likelihood that a victim of a narcissist will be invalidated, or told to work things out with their narcissistic partner, or worse, be blamed for the dysfunction in the relationship and told by their therapist that they need to work on their own "issues" and unresolved childhood "baggage." Sadly, this has been reported to me by my own patients more times than I can count because many of my patients who are victims and survivors of narcissistic abuse came to me after having previously seen multiple mental health professionals who unwittingly blamed them for their relationship problems with a narcissistic partner.

These patients desperately searched high and low for someone to truly understand the situation they were in, only to be told by several professionals that the real issue had little to do with their relationship or their partner. The "real" issue was that *they* were "codependent," or highly empathic, or insecurely attached, and if they just focused on themselves rather than on their partner, they would feel better.

Many mental health professionals don't actually "believe" that personality disorders exist as a distinct classification of disorders and instead conceptualize these disorders as environmentally learned behaviors with treatable and even curable symptoms. Many mental health professionals view personality pathology as a form of neurodivergence and promote this thinking error to victims, thereby invalidating them and inadvertently aligning with their narcissistic abuser. To reiterate one of the main points of this book, many mental health professionals see personality pathology as a result of complex trauma and believe personality disorders are caused by adverse childhood experiences and therefore should be theorized and treated as such. The emotional and physical harm this widespread erroneous belief has caused victims of narcissistic abuse and is continuing to cause is overwhelming.

Recovering from Narcissistic Abuse

If you are experiencing some or all of the symptoms and indicators of traumatic cognitive dissonance discussed throughout this chapter, it's important to remember that these are *symptoms*, not *personality traits*. These symptoms don't define you and are not the result of who you are; they are the result of what happened to you. Essentially, these symptoms do not make you like your abuser, and they can and will dissipate with the right kind of help. In a later chapter, I offer suggestions on finding the right kind of support for narcissistic abuse recovery.

Chapter Eleven
Narcissistic Children

The "Taboo" Topic

The idea that children are born with traits that predispose them to being narcissistic regardless of family upbringing or traumatic experiences is a vehemently debated topic for obvious reasons. No one wants to believe that some children are born with wiring that makes them prone to being antisocial and abusive throughout life. Discussions about kids and psychopathology are always controversial because nobody wants to believe that children are capable of committing heinous acts or being abusive. It is much more palatable to attribute children's antisocial or abusive behavior or disobedience in general to something related to the family dynamic or to the family system.

But the reality is that empirical research, including longitudinal studies which measure two or more different time points that span decades, reveal that children predisposed to have a personality disorder begin to exhibit stable (meaning consistent and inflexible) pathological personality traits and inappropriate behavior, including antisocial behavior, before the age of four.[183] Depending on their brain wiring from the start, some children are deficient in executive functions that enable them to have cognitive control over their thoughts, feelings, and behaviors.[184]

Because of this, some children, even preschool children, exhibit severe behavioral problems that are harmful and abusive to other children and even to authority figures, including disobedience, temper tantrums, lying, stealing, aggression, and violence.[185] These children lack the internal brake pedal that most other children have that enables them to weigh out consequences, including emotional consequences such as guilt, shame, or remorse. Empirical research has undeniably demonstrated

---

[183] R. D. Hare, *Without Conscience: The Disturbing World of the Psychopaths Among Us* The Guilford Press, (1993).; Paulina F. Kernberg et al., *Personality Disorders in Children and Adolescents* (Basic Books, 2000).; Anthony Petrosino et al., "Antisocial Behavior of Children and Adolescents: Harmful Treatments, Effective Interventions, and Novel Strategies," in *The Science and Pseudoscience in Clinical Psychology*, ed. Scott O. Lilienfeld, Steven J. Lynn, and Jeffrey M. Lohr, 2nd ed. (The Guilford Press, 2015), 500-532.

[184] Nadia M. Dias et al., "Can Executive Function Predict Behavior in Preschool Children?" *Psychology & Neuroscience* 10, no. 4 (2017): 383-393.

[185] Nadia M. Dias et al., "Can Executive Function Predict Behavior in Preschool Children?" *Psychology & Neuroscience* 10, no. 4 (2017): 383-393.; Anthony Petrosino et al., "Antisocial Behavior of Children and Adolescents," in *The Science and Pseudoscience in Clinical Psychology*, ed. Scott O. Lilienfeld, Steven J. Lynn, and Jeffrey M. Lohr, 2nd ed. (The Guilford Press, 2015), 500-532.

that some children are in fact narcissistic from the start.[186]

The truth is this: there are many exceptions to the "rule" that childhood adversity is the cause of maladaptive behaviors in children. There are some children who behave in ways that are inappropriate for their age and for their development. Children who are consistently and regularly disobedient, aggressive, violent, impulsive, uncooperative, and controlling are simply not best understood from the perspective of empathic failures or attachment ruptures.

Some children, whether we like to admit it or not, are naturally ruthless. Children who lie about very serious matters, including abuse; children who harm or kill the family pet; and children who sexually abuse others, exist. There are children who coerce siblings or peers who lack the mental capacity to consent or say no into sexual situations. There are children who viciously bully other people. Children who expect special treatment, disregard the feelings and needs of others, and don't

---

[186] Paulina F. Kernberg et al., *Personality Disorders in Children and Adolescents* (Basic Books, 2000).; Anthony Petrosino et al., "Antisocial Behavior of Children and Adolescents," in *The Science and Pseudoscience in Clinical Psychology*, ed. Scott O. Lilienfeld, Steven J. Lynn, and Jeffrey M. Lohr, 2nd ed. (The Guilford Press, 2015), 500-532.

express guilt, shame, or regret for the impact their behaviors and their words have on other people—including adults—are not "normal." Neuroimaging findings on the brains of even young children who meet the criteria for personality disorders like narcissism have shown the same neurobiological abnormalities discussed in chapter 8.

Many theorists and therapists have suggested that *all* children exhibit narcissism to some extent, but this kind of narcissism is a normal stage or normal course of development that the child will outgrow. However, longitudinal studies and empirical research have shown that some children do not outgrow their narcissistic traits and behaviors. Traits and behaviors cannot be constituted or conceptualized as age appropriate if they are not outgrown.

While it might be true that almost all children exhibit narcissistic *tendencies* early in life in the sense that they don't possess the brain wiring not to be self-centered, as they grow and learn how to share and care and love, children who have normal-functioning, adaptive brains outgrow their self-absorbed and self-centered tendencies. But not all children outgrow their narcissism. We still have to account for those children who do not outgrow maladaptive behaviors even in environments where social learning, warmth, nurturing, empathy, and love are readily available to them. Whether we like to admit it or not, some children are not like other children. Some children are hardwired with pathology in their personalities.

Pathological Narcissism in Childhood

Every child has a biologically-based temperament that begins to develop before they are even born. Some temperaments predispose some children to narcissism.[187] Environmental factors can enhance an already predisposed temperament, but the temperament is there no matter what due to the presence of neurobiological abnormalities that are not induced by the external environment.[188] This is not tragic news. It is good news, because children with personality disorders can be helped.[189] The sooner parents recognize the

---

[187] Paulina F. Kernberg et al., *Personality Disorders in Children and Adolescents* (Basic Books, 2000).

[188] Frederick L. Coolidge et al., "Heritability of Personality Disorders in Childhood: A Preliminary Investigation," *Journal of Personality Disorders* 15, no. 1 (2001): 33-40.; J. Kagan, *The Human Spark: The Science of Human Development*, Basic Books (2013); Ted Reichborn-Kjennerud, "The Genetic Epidemiology of Personality Disorders," *Dialogues in Clinical Neuroscience* 12, no. 1 (2010): 103-114.

[189] Edward Bleiberg, *Treating Personality Disorders in Children and Adolescents* (The Guilford Press, 2001).; Paulina F. Kernberg et al., *Personality Disorders in Children and Adolescents* (Basic Books, 2000).

indicators, the sooner the right interventions can begin.

Below are indicators of narcissism in early childhood to look for. Keep in mind, if these occur once in a while—no problem. We are looking for pervasive (meaning across the board in all areas of life) and persistent (meaning the rule rather than the exception).

Indicators of Narcissism in Early Childhood

- Requires constant admiration.
- Reacts with rage or emptiness when needs are not met.
- Frequently complains of boredom.
- Becomes entitled and exploitative when being nurtured.
- Cannot tolerate losing (and will end friendships over losing).
- Has environmental demands no one can realistically fulfill.
- Acts as though nothing is ever "enough."
- Is deficient in empathy.
- Violates generational boundaries.
- Is abusive to siblings.
- Does not express guilt, shame, or remorse after consequences of bad behavior.

- Does not modify behavior after consequences and blames circumstances and others for their behavior.
- Does not express gratitude or appreciation.
- Does not admit to dependency on adults.
- Has attachments that are superficial and lacking in warmth or sincere affection.
- Often lies and does not admit the truth even after there is evidence that they are lying.
- Deliberately creates the illusion of attachment to others to exploit and control rather than to bond or connect with vulnerability.
- Manages their image and impression to prevent humiliation and to ensure gratification.
- Gives orders rather than receives them, even to adults.

Treatment Recommendations for Narcissistic Children

Interventions that focus on increasing reflective capacities in children with narcissism have proven effective, so effective, in fact, that neuroimaging studies following treatment have revealed that brain functioning in these previously emotionally ruthless young patients has normalized.[190] Knowing that the cause of narcissism is a genetically fixed temperament is good news because we can do something about it. Cognitive behavioral therapy approaches have been developed to correct maladaptive thoughts in narcissistic children.[191] Cognitive behavioral techniques such as aggression management, social skills training, and moral reasoning have been shown to decrease antisocial behaviors and reduce repeated violations of authority in narcissistic children.[192]

Mentalization-based therapy has also been shown to be effective at reducing negative behaviors in narcissistic children and adolescents. This type of therapy involves the therapist guiding the patient toward developing a sense of self and learning how to practice mindfulness for improved regulation of behavior and impulse control.[193]

---

[190] Edward Bleiberg, *Treating Personality Disorders in Children and Adolescents* (The Guilford Press, 2001).

[191] Edward Bleiberg, *Treating Personality Disorders in Children and Adolescents* (The Guilford Press, 2001).; Anthony Petrosino et al., "Antisocial Behavior of Children and Adolescents," in *The Science and Pseudoscience in Clinical Psychology*, ed. Scott O. Lilienfeld, Steven J. Lynn, and Jeffrey M. Lohr, 2nd ed. (The Guilford Press, 2015), 500-532.

[192] Anthony Petrosino et al., "Antisocial Behavior of Children and Adolescents," in *The Science and Pseudoscience in Clinical Psychology*, ed. Scott O. Lilienfeld, Steven J. Lynn, and Jeffrey M. Lohr, 2nd ed. (The Guilford Press, 2015), 500-532.

[193] Anthony Petrosino et al., "Antisocial Behavior of Children and Adolescents," in *The Science and Pseudoscience in Clinical Psychology*, ed. Scott O. Lilienfeld, Steven J. Lynn, and Jeffrey M. Lohr, 2nd ed. (The Guilford Press, 2015), 500-532.

These treatment methods are evidenced based and accessible, which is good news. If you are a parent struggling with a child and the previously mentioned indicators seem to match your child's social functioning and behavior, consider looking into the above treatment methods for your child—and for your sanity! Untreated pathological narcissism does not go away on its own and in fact gets much worse, so the sooner the interventions start, the better the outcome.[194]

A Word of Caution about Choosing a Therapist

Therapists don't all practice the same way, and because of this, some therapists misdiagnose and mistreat patients in ways that can cause serious harm. The treatment of personality disorders—even the existence of them in children and adolescents—has been contested or even outright denied by many therapists. This is a form of malpractice as far as I'm concerned.

I have included several references in the back of this book that I recommend for further reading

---

[194] Paulina F. Kernberg et al., *Personality Disorders in Children and Adolescents* (Basic Books, 2000). ; Stuart C. Yudofsky, *Fatal Flaws* (American Psychiatric Association Publishing, 2005).

regarding the treatment of personality disorders in children and adolescents. When searching for a therapist for your child, make certain you confirm that the therapist both treats personality disorders and has experience treating them in children and teenagers.

Chapter Twelve
Treatment for Narcissistic Adults

The Difficulty of Treating Narcissistic Personality Disorder

Narcissistic individuals present as challenges in therapy just as they do in everyday life. As stated before, narcissists tend to seek treatment only when their life has failed to produce the desired results and expectations they hold for themselves no matter how far-fetched or unrealistic those expectations are. Their personality trait pathology limits their insight, so they typically deny any responsibility for their circumstances. Instead, they externalize blame onto others as the reason why life isn't going the way they want it to go. They present as helpless victims of life's circumstances rather than as active participants in their own problems and shortcomings. Most present as vulnerable and misunderstood, which can throw a therapist off and cause them to default to a person-centered or collaborative therapy approach, which is not effective when working with narcissism.

While in therapy, narcissists typically complain about relationships with family, friends, employers, and rules and legal standards that they refuse to abide by or accept.[195] Because they don't present themselves authentically, it is all too common for mental health providers to overlook a diagnosis of narcissistic personality disorder in the initial treatment session and in the early stages of treatment in general. Comorbid conditions such as depression or anxiety make overlooking narcissistic pathology even more likely.[196]

Patients with narcissistic personality disorder are difficult to interact with in a professional therapeutic setting, just as they are difficult to interact with in personal relationships. Mutually agreed-upon goals and a mutually agreed-upon treatment frame are essential from the outset.[197] Even if a therapeutic relationship can be commenced, it is to be expected that a narcissistic patient will not

---

[195] Jessica Yakeley, "Current Understanding of Narcissism and Narcissistic Personality Disorder," *BJPsych Advances* 24 (2018): 305-315.

[196] David Kealy, Gail A. Hadjipavlou, and John S. Ogrodniczuk, "Moving the Field Forward: Commentary for the Special Series 'Narcissistic Personality Disorder – New Perspectives on Diagnosis and Treatment'," *Personality Disorders: Theory, Research, and Treatment* 5, no. 4 (2014): 444-445.

[197] Jessica Yakeley, "Current Understanding of Narcissism and Narcissistic Personality Disorder," *BJPsych Advances* 24 (2018): 305-315.

cooperate and collaborate throughout treatment.[198] Furthermore, recurrent setbacks and ruptures are to be expected throughout the therapeutic process.[199]

Sometimes, narcissists will play nice in therapy as they do when they are using flattery and charm in personal interactions because they are good at knowing what is expected of them even when no therapeutic change has occurred in reality.[200] This is another covert tactic that throws many therapists off—everything seems to be going well when nothing is actually going on at all. Narcissists eventually become bored with this pretend game of theirs and will abandon therapy prematurely if the therapist is unaware of what is happening. This makes the therapist's job a balancing act because when the narcissist's *pretend-nice* persona is called out by an astute therapist for what it is—manipulation—narcissists tend to become enraged and drop out of therapy.

---

[198] Gregory W. Lester, *Borderline, Narcissistic, Antisocial, and Histrionic Personality Disorders* (PESI Publishing, 2024).

[199] Jessica Yakeley, "Current Understanding of Narcissism and Narcissistic Personality Disorder," *BJPsych Advances* 24 (2018): 305-315.

[200] Jessica Yakeley, "Current Understanding of Narcissism and Narcissistic Personality Disorder," *BJPsych Advances* 24 (2018): 305-315.

Transference-Focused Therapy

Countertransference reactions in therapy are very useful in identifying narcissism.[201] Countertransference refers to the internal emotional reaction of the therapist to a patient in session. Common countertransference reactions to narcissists that therapists experience include feelings of incompetence and impatience, feeling devalued and exploited, and feeling fearful and inhibited, among other reactions.[202] These reactions in the therapist are to be understood as informative, helpful, and even necessary rather than as inappropriate.

Transference-focused therapy has proven to be effective in treating narcissistic personality disorder.[203] Although originally developed to treat borderline personality disorder, transference-focused therapy is a supportive approach that confronts the here-and-now dynamics of

---

[201] Aaron L. Pincus et al. "Narcissistic Grandiosity and Narcissistic Vulnerability in Psychotherapy," *Personality Disorders* 5, no. 4 (2014): 439-443.

[202] Aaron L. Pincus et al. "Narcissistic Grandiosity and Narcissistic Vulnerability in Psychotherapy," *Personality Disorders* 5, no. 4 (2014): 439-443.

[203] David Kealy, Gail A. Hadjipavlou, and John S. Ogrodniczuk, "Therapists' Perspectives on Optimal Treatment for Pathological Narcissism," *Personality Disorders: Theory, Research, and Treatment* 8, no. 1 (2017): 35-45.; Jeremy Safran, "Psychoanalytic Therapy Process," in *Psychotherapy Theories and Techniques: A Reader*, ed. Gary R. Vandenbos, Edward Meidenbauer, and Julia Frank-McNeil (American Psychological Association, 2014), 281-288.

the narcissistic patient's way of relating to the therapist.[204] Therapists assist narcissistic patients in making connections as to what is happening within the session. This is achieved by the therapist making cause-and-effect statements and observations of how the narcissist is behaving and interacting in the therapy session—remarks that would otherwise seem inappropriate or intrusive in non-professional social settings.[205]

This type of therapy also involves pointing out to the narcissistic patient where their problem-solving capabilities fall short and modeling and encouraging new ways of thinking, feeling, and behaving. The narcissist then is encouraged to practice these new ways of thinking, feeling, and behaving during therapy sessions with the therapist.[206] By drawing attention to the way the narcissistic patient behaves and perceives reality, transference-focused therapy offers the rare opportunity for the narcissistic patient to observe

---

[204] Jeremy Safran, "Psychoanalytic Therapy Process," in *Psychotherapy Theories and Techniques: A Reader*, ed. Gary R. Vandenbos, Edward Meidenbauer, and Julia Frank-McNeil (American Psychological Association, 2014), 281-288.

[205] Gregory W. Lester, *Borderline, Narcissistic, Antisocial, and Histrionic Personality Disorders* (PESI Publishing, 2024).

[206] Jessica Yakeley, "Current Understanding of Narcissism and Narcissistic Personality Disorder," *BJPsych Advances* 24 (2018): 305-315.

themselves in the here and now, right in the midst of the process of experience.[207]

Mentalization-Based Therapy

Mentalization-based therapy is also an effective treatment method for adults who are pathologically narcissistic.[208] By assisting the patient in developing the capacity to reflect on their own state of mind and the states of mind of others, this type of therapeutic intervention can work to reduce maladaptive behaviors and dramatic escalations and replace them with mindful problem-solving skills.[209]

A caveat: both transference-focused therapy and mentalization-based therapy are long-term treatment interventions and are not appropriate for all narcissists—only the ones who are highly treatable and not physically dangerous.

---

[207] Jeremy Safran, "Psychoanalytic Therapy Process," in *Psychotherapy Theories and Techniques: A Reader*, ed. Gary R. Vandenbos, Edward Meidenbauer, and Julia Frank-McNeil (American Psychological Association, 2014), 281-288.

[208] Jessica Yakeley, "Current Understanding of Narcissism and Narcissistic Personality Disorder," *BJPsych Advances* 24 (2018): 305-315.

[209] Gregory W. Lester, *Personality Disorders: Advanced Diagnosis, Treatment & Management* (PESI Publishing, 2018).
Jessica Yakeley, "Current Understanding of Narcissism and Narcissistic Personality Disorder," *BJPsych Advances* 24 (2018): 305-315.

Schema-Focused Therapy

Schema-focused therapy looks to identify generalized automatic thoughts and beliefs that are maladaptive, and replace those generalizations with mature, adult-like thoughts and beliefs that improve psychological functioning.[210] Therapists use schema-based interventions to help narcissistic patients re-evaluate the legitimacy and practicality of their maladaptive schemas in order to start the process of formulating more realistic beliefs.[211] This in turn can work to increase interpersonal functioning and possibly cultivate the development of a cognitive understanding of empathy in some, but not all, narcissistic individuals.[212]

This type of therapeutic method also focuses more on the patient's past and their relationships outside of therapy rather than on the present, as opposed to transference-focused therapy

[210] Lawrence P. Riso and Christopher McBride, "Schema Therapy," in *Psychotherapy Theories and Techniques: A Reader*, ed. Gary R. Vandenbos, Edward Meidenbauer, and Julia Frank-McNeil (American Psychological Association, 2014), 345-350.

[211] Lawrence P. Riso, Robert E. Maddux, and Nicole T. Santorelli, "Schema Therapy Process," in *Psychotherapy Theories and Techniques: A Reader*, ed. Gary R. Vandenbos, Edward Meidenbauer, and Julia Frank-McNeil (American Psychological Association, 2014), 351-355.

[212] Jessica Yakeley, "Current Understanding of Narcissism and Narcissistic Personality Disorder," *BJPsych Advances* 24 (2018): 305-315.

and mentalization-based therapy which focus on the here and now.[213] Schema-focused therapy is utilized with narcissistic patients who are moderately treatable.[214]

Additional Treatment Methods for Narcissism

In addition to the therapeutic methods above, brief strategic and structural therapies can be used to treat narcissism when time is limited and the patient is not very treatable. These types of therapy interventions focus on trying to improve problems one at a time in a shorter amount of time.[215] A specific behavior or issue can be addressed that is of concern, and the treatment involves strategic focus on that one problem until change occurs.[216]

Experts in personality disorder treatment have effectively treated many narcissistic

---

[213] Gregory W. Lester, *Borderline, Narcissistic, Antisocial, and Histrionic Personality Disorders* (PESI Publishing, 2024).

[214] Gregory W. Lester, *Personality Disorders: Advanced Diagnosis, Treatment & Management* (PESI Publishing, 2018).

[215] Gregory W. Lester, *Borderline, Narcissistic, Antisocial, and Histrionic Personality Disorders* (PESI Publishing, 2024).

[216] Alan Godwin and Gregory W. Lester, *Demystifying Personality Disorders* (PESI Publishing, 2021).

patients,[217] but in order for any treatment intervention to work with a personality pathology like narcissism, an individual must desire change, believe change is possible, know that change is necessary to some degree, and practice new ways of behaving to create new habits.[218] Ideally, treating narcissism successfully will result in the narcissistic individual learning how to engage with others, collaborate, tolerate constructive feedback from others, share with others, appreciate others, be aware of the feelings of others, and be happy for others when they succeed and are doing well.[219]

In terms of self-help for victims and survivors of narcissistic abuse who are looking for answers for themselves about their loved ones, resources like this book, personal therapy, as well as community support, can be very beneficial.[220]

---

[217] Alan Godwin and Gregory W. Lester, *Demystifying Personality Disorders* (PESI Publishing, 2021).

[218] Christopher E. Sleep, Donald R. Lynam, and Joshua D. Miller, "Understanding Individuals' Desire for Change, Perceptions of Impairment, Benefits, and Barriers of Change for Pathological Personality Traits," *Personality Disorders: Theory, Research, and Treatment* 13, no. 3 (2021): 245-253.

[219] Gregory W. Lester, *Borderline, Narcissistic, Antisocial, and Histrionic Personality Disorders* (PESI Publishing, 2024).

[220] Stuart C. Yudofsky, *Fatal Flaws* (American Psychiatric Association Publishing, 2005).

A Word of Caution When Seeking a Therapist

If you are seeking a mental healthcare provider, either privately or through your health insurance company, it is extremely important to understand that there is no such thing as a one-size-fits-all therapy. Therapeutic specialties and treatment methods vary so widely that it is crucial that patients seeking mental health care understand just how different one therapist can be from another. In order to make an informed decision, the prospective patient wants a therapist who can demonstrate through not only their credentials but also their professional experience and specialties that they can successfully treat the problems the patient is coming to see them for.

I cannot emphasize this enough. There are so many occasions when a therapist who is not well informed in a particular treatment approach attempts to treat a certain mental health condition or disorder and misdiagnoses a patient and/or misses the appropriate diagnosis altogether.[221] There are also occasions when therapists make recommendations that can be hazardous to the mental and physical health of their patients.

---

[221] Stuart C. Yudofsky, *Fatal Flaws* (American Psychiatric Association Publishing, 2005).

Unfortunately, some therapists believe they are well-versed in understanding all treatment methods and interventions when, in fact, they are not.

My Personal Approach as a Therapist

Once a patient—after having been referred to me or having reached out to me on their own—has decided that the ways in which I work and my specialties are in alignment with the treatment they are seeking, we do a consultation. The consultation allows me to see if I am the right fit for the person, as I am not a good fit for everyone even if I have the credentials and the competency to treat them.

It is very important that therapists do not work with patients who present in ways that are outside of the scope of their competence. There is a difference between the scope of practice and the scope of competence. For example, I am a licensed psychotherapist who specializes in the treatment of trauma. My scope of competence includes childhood, adult, single-event, and cumulative trauma as well as pathological relationship abuse. Successfully treating trauma is literally what I do all day. I also have another specialty, which is the treatment of individuals who meet the criteria for cluster B personality disorders, including narcissistic personality disorder. Therefore, I have a scope of competence in two significant areas.

I have been successful in treating victims and survivors of the abuse that occurs in relationships with cluster B personalities, as well as treating cluster B personalities themselves. The ways in which these populations are treated are drastically different. You do not use the same interventions for treating trauma as you do for treating narcissistic personality disorder, and there is a reason for this.

When I am working with patients who meet the criteria for a cluster B personality disorder, like narcissism, for example, the patients must agree to stay within the designated treatment frame that is established and discussed prior to the commencement of treatment. This designated treatment frame is much stricter and more rigid than the typical parameters that are agreed upon when working with a patient who is not personality disordered. If a personality disordered patient can't or won't abide by the agreed-upon parameters of the treatment frame, I tell them they must seek help elsewhere. To not recognize personality pathology and to further not have an established treatment frame from the moment therapy begins guarantees failure when treating a narcissist. And this, sadly, happens more often than it should in clinical practice.

We cannot all know everything. That is why we specialize and stick to our specialties. For example, if a patient is struggling with a substance-use disorder, I am the wrong therapist to see because I do not specialize in treating that population or those types of disorders. If an individual is suffering from schizophrenia, I am the wrong therapist to see because I do not specialize in the treatment of psychotic disorders.

Too often those in the general public assume that all therapists know how to treat all mental disorders and conditions and that therapists typically treat patients in the same or similar ways. This is categorically untrue. This belief can be harmful for patients seeking therapy, even though it's no fault of their own that they believe it. We can't all be proficient in everything. I have no business being in the room with somebody who meets the criteria for schizoaffective disorder. But if you put me in the room with somebody who has a history of sexual abuse or emotional neglect, or who has been the victim of severe manipulation by a narcissist, I'm the right therapist for the job.

Personal Experience Is Not Expertise

There is a myth that prevails in popular culture that if you've been through it, you know exactly what to do about it. I don't mean any disrespect to anybody who's been through hell. In fact, I personally have been through hell and have survived a highly dangerous and abusive relationship with a cluster B personality from my past. But going through hell does not mean you have even the slightest clue how to be in a room with a patient and successfully intervene in ways that can effectively alleviate and reduce their cognitive and emotional struggles. Even if the experiences and circumstances a patient shares with you are almost identical to your own, it is not true that personal experience makes you a professional.

For example, in the beginning pages of this book I shared the story of my maternal grandmother; she is someone who meets all the criteria for not one but two personality disorders. But it was not until I earned a doctorate in psychology and spent thousands of hours working in clinical settings under clinical supervision, absorbed the most current research, participated in clinical consultations, and embarked upon a prolonged course of continuing education and specialized training, that I could ethically consider myself proficient to treat somebody who presents with the same disorders as my grandmother.

Having a grandmother who met the criteria for narcissistic personality disorder and borderline personality disorder has nothing to do with my professional ability to help others. My personal experience of witnessing personality pathology in my own family might lend itself to a bit of an unconscious familiarity with the drama and dysfunctional dance of narcissistic relationships, but it does not qualify me to understand how to intervene on behalf of patients who are like her or who know someone like her.

The way we intervene and resolve the problems in other people's lives has nothing to do with the way we have solved problems in our own lives. If this were the case, then anybody who is in a marriage can say that they know everything about how to have a successful marriage. This contradicts the fact of the high divorce rate in our society; if being in a successful marriage is the prerequisite for being an expert in marriage, the divorce rate should be nonexistent.

Why go to rehab if you're an alcoholic if being an alcoholic is the prerequisite for learning how to successfully get sober and get other individuals to stop drinking? Do you see how this doesn't make sense? How you have dealt with or have successfully overcome your personal challenges really doesn't hold much weight when you are trying to help somebody else overcome their personal challenges. This is especially true when dealing with narcissists simply because interventions that work with narcissists aren't all that relatable.

In mental health care, there is a difference between teaching somebody what you know and teaching them what they don't know and need to know. And what you know based on your personal experience might not be the slightest bit helpful to somebody who needs to learn something about themselves. Knowing how to treat mental health conditions, especially ones like personality disorders, is way more demanding and challenging than it may seem.

I don't mean any offense by this. I know this is a sensitive subject. But I am including myself in this discussion. I do not have any business treating personality disorders because I know firsthand what it's like to deal with a personality-disordered family member. On the other hand, I do have the right to treat personality disorders because of my professional training and qualifications.

Experience doesn't qualify you to be an expert. Training does.

One final note: please take good care of yourself, your relationships, and your children by seeking professional help from qualified specialists who can prove their competency and expertise with professional qualifications.

Chapter Thirteen
In Summation: The Narcissist at Large

Summing Up Narcissistic Personality Disorder

Narcissists are present in every culture, in every society, economic class, and professional occupation. Narcissists vary in intelligence, talent, skill, and capacity just like everyone else, regardless of parenting style or family upbringing.

In consideration of this, it makes the most sense to view narcissistic personality disorder as a trait-related disorder rather than a trauma-related disorder. Narcissists have a severe deficiency in traits such as equality, empathy, agreeableness, and conscientiousness, among others. Narcissists have an excess of traits such as grandiosity, arrogance, entitlement, combativeness, antagonism, and callousness, among others.

Narcissists have a biologically based temperament with fixed trait levels that are persistent and pervasive rather than flexible and malleable. Genetic factors demonstrate that these pathological levels of traits are highly heritable. Narcissists exhibit such things as missing or immature neurons in certain brain regions, faulty wiring, deficits in matter and volume in certain brain regions, and other neurobiological impairments.

Non-trauma-related environmental risk factors that interact with genes work to sustain or even exacerbate the innate pathological trait variations in the narcissist. It also seems that narcissistic individuals tend to gravitate toward environments suitable for the expression of the above trait deficiencies and excesses, thereby exacerbating the predisposition by playing them out in environments of their own choosing and further hardening the neural connections.

It has also been shown that these pathological traits are evident in preschool-aged children and contribute to childhood behavioral problems regardless of parental style or family environment. Childhood abuse does not seem to be a significant risk factor or predictor of narcissistic personality disorder despite the vast amount of clinical literature that has suggested this for over a century. Control groups and longitudinal studies, including decades-long follow-up studies, as well as family, twin, and adoption studies, don't reveal a correlation between childhood abuse and narcissistic personality disorder as originally thought.

Where We Are Now

Now that we have technologies that can actually see the brain and monitor and record brain functioning and activity in real time, we know a lot of things that previous generations of psychiatrists, psychologists, and therapists could not have possibly known. What psychoanalytic, psychodynamic, social learning, and attachment theories didn't take into consideration previously when viewing childhood adversity and trauma is now being explored, and new conclusions are being drawn about narcissistic personality disorder. There now exists hard data that shows inherited and varying levels of innate resiliency in each individual personality. Our personalities—once thought to be the product of our early environmental experiences—are very much shaped by our genes.[222]

We are not born as blank slates, completely at the mercy of our parents' personalities, parenting behaviors, and extended social environment.[223] Children raised in the same household by the

---

[222] Guo Ma et al., "Genetic and Neuroimaging Features of Personality Disorders," *Neuroscience Bulletin* 32, no. 3 (2016): 286-306.
[223] Joel Paris, *Myths of Childhood* (Routledge, 2014).

same parents or caregivers differ in temperament and environmental sensitivity from birth.[224] Anyone who grew up with a biological sibling knows that individual experiences of family dynamics and characteristics are not anywhere near identical among siblings, and siblings have very different relationships with and experiences of their parents.[225]

Social and environmental stressors, and responses to them, are directly related to our individual, innate temperamental resilience, and innate temperamental resilience is the result of genetic factors.[226] Two children who experience the same parental behavior in the same household at the same time will experience and internalize that behavior completely differently from one another.[227]

In short, science has proven that personality traits that influence differences in the perception of parenting are inherited.[228] Does environment matter? Absolutely. But not so much when it comes

---

[224] Joel Paris, *Myths of Childhood* (Routledge, 2014).

[225] Susan C. South and Nicholas J. DeYoung, "The Remaining Road to Classifying Personality Pathology in the *DSM*-5: What Behavior Genetics Can Add," *Personality Disorders: Theory, Research, and Treatment* 4, no. 3 (2013): 291-292.

[226] Joel Paris, *Myths of Trauma* (2023).; Ted Reichborn-Kjennerud, "The Genetic Epidemiology of Personality Disorders," Dialogues in Clinical Neuroscience 12, no. 1 (2010): 103-114.

[227] Joel Paris, *Myths of Childhood* (Routledge, 2014).

[228] Joel Paris, *Myths of Childhood* (Routledge, 2014).

to narcissism.[229] Young children are not passive, submissive, reflexive, lifeless recipients of parenting styles, whether the parenting style is "good" or "bad."[230] Children have their own minds. They are not blank slates. And finally, genes affect normalcy just as much as they affect disorder.[231]

Afterword

This book is not intended to be an extensive treatment manual or how-to guide, but my hope is that understanding normal and abnormal personality and accepting that some people do the things they do because they are wired to do so will assist you, the reader, in understanding the oft-misinterpreted narcissist. I further hope that if you are dealing with someone with narcissistic personality disorder—which is often at best, challenging, at worst, abusive—you will now be able to make informed decisions that are in *your* best interest. If we presume trauma is the root cause of narcissism, we can get stuck in a role such as caretaker, rescuer, or victim, over and over again. If we see

---

[229] Gregory W. Lester, *Borderline, Narcissistic, Antisocial, and Histrionic Personality Disorders* (PESI Publishing, 2024).
[230] Joel Paris, 2014. *Myths of Childhood* (Routledge, 2014).
[231] Siddhartha Mukherjee, *The Gene* (Scribner, 2017).

narcissism for what it truly is, we can adjust our expectations and decide for ourselves what level of involvement we want to have with people who are pathologically narcissistic.

We need to respect ourselves by developing and maintaining parameters for how close we allow people to get to us who are troublesome, abusive, manipulative, and violent. The benefit of the doubt has to be earned by others and not freely given. If someone violates our personal boundaries, we need to be willing to deliver a consequence that will either motivate them to change for good or motivate us to modify our proximity to them. It's up to us to be the guardians and caretakers of our hearts and minds. To leave this up to a narcissist in our lives is not the best idea.

References

Baker, Laura A. "The Biology of Relationships: What Behavioral Genetics Tells Us about Interactions among Family Members." *De Paul Law Review* 56, no. 3 (2007): 837-846.

Barrett, Lisa Feldman. *Seven and a Half Lessons about the Brain*. Mariner Books. (2021).

Beam, Cassandra R., Paola Pezzoli, Jane Mendle, S. Alexandra Burt, Michael C. Neale, Steven M. Boker, Pamela K. Keel, and Kelly L. Klump. "How Nonshared Environmental Factors Come to Correlate with Heredity." *Developmental Psychopathology* 34, no. 1 (2022): 321-333.

Beck, Aaron T., and Arthur Freeman, eds. *Cognitive Therapy of Personality Disorders*. The Guilford Press (1990).

Bleiberg, Edward. *Treating Personality Disorders in Children and Adolescents: A Relational Approach*. The Guilford Press (2001).

Brown, Stephanie L., Christine Paradise, and Barbara Brennan. *Intensive Training on Narcissistic and Psychopathic Abuse*. PESI, Inc. (2021).

Brummelman, Eddie, Ceren Gurel, Sander Thomaes, and Constantine Sedikides. "What Separates Narcissism from Self-Esteem? A Social-Cognitive Perspective." In *Handbook of Trait Narcissism: Key Advances, Research Methods, and Controversies*, edited by Anthony D. Hermann, Amy B. Brunell, and Joshua D. Foster, 47-55. Springer International Publishing (2018).

Coolidge, Frederick L., Lezlee L. Thede, and Kerry L. Jang. "Heritability of Personality Disorders in Childhood: A Preliminary Investigation." *Journal of Personality Disorders* 15, no. 1 (2001): 33-40.

Di Sarno, Marco, Renato Di Pierro, and Fabio Madeddu. "The Relevance of Neuroscience for the Investigation of Narcissism: A Review of Current Studies." *Clinical Neuropsychiatry* 15, no. 4 (2018): 242.

Dias, Nadia M., Bruna T. Trevisan, Camila B. Leon, Ana P. Prust, and Alessandra G. Seabra. "Can Executive Function Predict Behavior in Preschool Children?" *Psychology & Neuroscience* 10, no. 4 (2017): 383-393.

Distel, Marijn A., Timothy J. Trull, Gonneke Willemsen, Jacqueline M. Vink, Christine A.

Derom, Michael Lynskey, Nicholas G. Martin, and Dorret I. Boomsma. "The Five-Factor Model of Personality and Borderline Personality Disorder: A Genetic Analysis of Comorbidity." *Biological Psychiatry* 66, no. 11 (2009): 1131-1138.

Elliot, A. *Psychoanalytic Theory: An Introduction.* Duke University Press (2000).

Fjelstad, Margalis. *Stop Caretaking the Borderline or Narcissist: How to End the Drama and Get on with Your Life.* Rowman & Littlefield Publishers, Inc. (2013).

Fontaine, Nathalie, and Essi Viding. "Genetics of Personality Disorders." *Psychiatry* 7, no. 3 (2008): 137-141.

Frankland, Adam G. *The Little Psychotherapy Book: Object Relations in Practice.* Oxford University Press, Inc. (2010).

Freud, Sigmund. "On Narcissism: An Introduction." In *The Freud Reader*, edited by Peter Gay, 545-562. Norton (1995).

Friedel, Robert O., Christine Schmahl, and Marijn Distel. "The Neurobiological Basis of Borderline Personality Disorder." In *Neurobiology of Personality Disorders*, edited by Christine

Schmahl, K. Luan Phan, Robert O. Friedel, and Larry J. Siever, 279-317. Oxford University Press (2018).

Gay, Peter, ed. *The Freud Reader*. Norton (1995).

Genetic Alliance, Understanding Genetics: A New York, Mid-Atlantic Guide for Patients and Health Professionals (2009) https://pubmed.ncbi.nlm.nih.gov/23304754/.

Gentry, J.E. *Forward-Facing Freedom: Healing the Past, Transforming the Present, A Future on Purpose*. Parker, CO: Outskirts Press (2021).

George, Frank R., and Derek Short. "The Cognitive Neuroscience of Narcissism." *Science Publications* (2017): 1-14.

Gerson, M.J. *Child Abuse and Trauma*. Westlake Village, CA: Institute of Advanced Psychological Studies (2021).

Godwin, Alan, and Gregory W. Lester. *Demystifying Personality Disorders: Clinical Skills for Working with Drama and Manipulation*. PESI Publishing (2021).

Green, Amanda, Rebecca MacLean, and Katherine Charles. "Female Narcissism: Assessment, Aetiology, and Behavioural Manifestations." *Psychological Reports* 125, no. 6 (2022): 2833-2864.

Hare, Robert D. *Without Conscience: The Disturbing World of the Psychopaths Among Us.* The Guilford Press (1993).

Hare, Robert D. *Hare PCL-R,* 2nd ed. Multi-Health Systems, Inc. (2003).

Holtzman, Nicholas S. "Did Narcissism Evolve?" In *Handbook of Trait Narcissism: Key Advances, Research Methods, and Controversies,* edited by Anthony D. Hermann, Amy B. Brunell, and Joshua D. Foster, 173-181. Springer International Publishing (2018).

Jauk, Emanuel, Christine Blum, Markus Hildebrandt, Kathrin Lehmann, Lara Maliske, and Philipp Kanske. "Psychological and Neural Correlates of Social Affect and Cognition in Narcissism: A Multimethod Study of Self-Reported Traits, Experiential States, and Behavioral and Brain Indicators." *Personality Disorders: Theory, Research, and Treatment* 15, no. 2 (2024): 157-171.

Jauk, Emanuel, and Philipp Kanske. "Can Neuroscience Help to Understand Narcissism? A Systematic Review of an Emerging Field." *Personal Neuroscience* (2021). https://doi.org/10.1017/pen.2021.1.

Kagan, J. *The Human Spark: The Science of Human Development.* Basic Books (2013).

Kealy, David, Gail A. Hadjipavlou, and John S. Ogrodniczuk. "Therapists' Perspectives on Optimal Treatment for Pathological Narcissism." *Personality Disorders: Theory, Research, and Treatment* 8, no. 1 (2017): 35-45.

Kealy, David, Gail A. Hadjipavlou, and John S. Ogrodniczuk. "Moving the Field Forward: Commentary for the Special Series 'Narcissistic Personality Disorder – New Perspectives on Diagnosis and Treatment'." *Personality Disorders: Theory, Research, and Treatment* 5, no. 4 (2014): 444-445.

Kernberg, Otto F. *Internal World and External Reality.* Aronson (1980).

Kernberg, Paulina F., Alan S. Weiner, and Karen K. Bardenstein. *Personality Disorders in Children and Adolescents.* Basic Books (2000).

Knipe, J. *EMDR Toolbox: Theory and Treatment of Complex PTSD and Dissociation: Theory and Treatment of Complex PTSD and Dissociation.* New York, NY: Springer Publications (2018).

Lester, Gregory W. *Personality Disorders: Advanced Diagnosis, Treatment & Management.* PESI Publishing (2018).

Lester, Gregory W. *Borderline, Narcissistic, Antisocial, and Histrionic Personality Disorders: Effective Treatments for Challenging Clients.* PESI Publishing (2024).

Lilienfeld, Scott O., Steven J. Lynn, and Jeffrey M. Lohr, eds. *Science and Pseudoscience in Clinical Psychology*, 2nd ed. The Guilford Press (2015).

Livesley, W. John, and Kerry L. Jang. "The Behavioral Genetics of Personality Disorder." *Annual Review of Clinical Psychology* 4 (2008): 247-274.

Luo, Yilu, and Huajian Cai. "The Etiology of Narcissism: A Review of Behavioral Genetic Studies." In *Handbook of Trait Narcissism: Key Advances, Research Methods, and Controversies*, edited by Anthony D. Hermann, Amy B. Brunell, and Joshua D. Foster, 149-156. Springer International Publishing (2018).

Luo, Yilu, Huajian Cai, and Huiwen Song. "A Behavioral Genetic Study of Intrapersonal and Interpersonal Dimensions of Narcissism." *Plos One* 9, no. 4 (2014). https://doi.org/10.1371/journal.pone.0093403.

Ma, Guo, Haofan Fan, Chao Shen, and Wei Wang. "Genetic and Neuroimaging Features of Personality Disorders: State of the Art." *Neuroscience Bulletin* 32, no. 3 (2016): 286-306.

Marlow-MaCoy, Ariel. *The Gaslighting Recovery Workbook: Healing from Emotional Abuse.* Callisto Media (2020).

Mason, Paul T., and Randi Kreger. *Stop Walking on Eggshells: Taking Your Life Back When Someone You Care About Has Borderline Personality Disorder,* 3rd ed., rev. New Harbinger Publications, Inc. (2020).

McGue, Matt. "Behavioral Genetics." Video file, February 27 2024. https://www.coursera.org/learn/behavioralgenetics.

Miles, Gabriel J., and Andrew J. Francis. "Narcissism: Is Parenting Style to Blame, or Is Their X-Chromosome Involvement?" *Psychiatry Research* 219 (2014): 712-713.

Miller, Joshua D., Donald R. Lynam, Courtney S. Hyatt, and W. Keith Campbell. "Controversies in Narcissism." *Annual Review of Clinical Psychology* 13 (2017): 291-315.

Miller, Joshua D., Thomas A. Widiger, and W. Keith Campbell. "Vulnerable Narcissism: Commentary for the Special Series 'Narcissistic Personality Disorder – New Perspectives on Diagnosis and Treatment'." *Personality Disorders: Theory, Research, and Treatment* 5, no. 4 (2014): 450-451.

Moran, Mark. "Understanding Personality Disorders Requires a New Way of Thinking." *Psychiatric News*, July 7, 2006. https://doi.org/10.1176/pn.41.13.0026.

Mukherjee, Siddhartha. *The Gene: An Intimate History*. Scribner (2017).

Paris, Joel. *Myths of Childhood*. Routledge (2014).

Paris, Joel. *A Concise Guide to Personality Disorders*. American Psychological Association (2015).

Paris, Joel. *The Intelligent Clinician's Guide to the DSM-5*. Oxford University Press (2015).

Paris Joel. *Treatment of borderline personality disorder: A guide to evidence-based practice*. The Guilford Press (2020).

Paris, Joel. *Myths of Trauma: Why Adversity Does Not Necessarily Make Us Sick*. Oxford University Press (2023).

Paris, Joel. *Fads and Fallacies in Psychiatry*. Cambridge University Press (2023).

Petrosino, Anthony, Patricia MacDougall, Melissa E. Hollis-Peel, Tamara A. Fronius, and Sara Guckenburg. "Antisocial Behavior of Children and Adolescents: Harmful Treatments, Effective Interventions, and Novel Strategies." In *The Science and Pseudoscience in Clinical Psychology*, edited by Scott O. Lilienfeld, Steven J. Lynn, and Jeffrey M. Lohr, 2nd ed., 500-532. The Guilford Press (2015).

Pincus, Aaron L., Nicole M. Cain, and Aidan G. Wright. "Narcissistic Grandiosity and Narcissistic Vulnerability in Psychotherapy." *Personality Disorders: Theory, Research, and Treatment* 5, no. 4 (2014): 439-443.

Reichborn-Kjennerud, Ted. "The Genetic Epidemiology of Personality Disorders." *Dialogues in Clinical Neuroscience* 12, no. 1 (2010): 103-114.

Reichborn-Kjennerud, Ted, and Kenneth S. Kendler. "Genetics of Personality Disorders." In *Neurobiology of Personality Disorders*, edited by Christine Schmahl, K. Luan Phan, Robert O. Friedel, and Larry J. Siever, 57-73. Oxford University Press (2018).

Resick, Patricia A., Candice M. Monson, and Kathleen M. Chard. *Cognitive Processing Therapy for PTSD: A Comprehensive Manual.* The Guilford Press (2017).

Riso, Lawrence P., Robert E. Maddux, and Nicole T. Santorelli. "Schema Therapy Process." In *Psychotherapy Theories and Techniques: A Reader*, edited by Gary R. Vandenbos, Edward Meidenbauer, and Julia Frank-McNeil, 351-355. American Psychological Association (2014).

Riso, Lawrence P., and Christopher McBride. "Schema Therapy." In *Psychotherapy Theories and Techniques: A Reader*, edited by Gary R. Vandenbos, Edward Meidenbauer, and Julia Frank-McNeil, 345-350. American Psychological Association (2014).

Robinson, Donald J. *Disordered Personalities*, 3rd ed. Rapid Psychler Press (2005).

Roth, Kimberlee, and Freda B. Friedman. *Surviving a Borderline Parent: How to Heal Your Childhood Wounds & Build Trust, Boundaries, and*

Self-Esteem. New Harbinger Publications, Inc. (2003).

Russell, Gary A. "Narcissism and the Narcissistic Personality Disorder: A Comparison of the Theories of Kernberg and Kohut." *British Journal of Medical Psychology* 58 (1985): 137-148.

Safran, Jeremy. "Psychoanalytic Therapy Process." In *Psychotherapy Theories and Techniques: A Reader,* edited by Gary R. Vandenbos, Edward Meidenbauer, and Julia Frank-McNeil, 281-288. American Psychological Association (2014).

Sanchez-Roige, Sandra, James C. Gray, John K. MacKillop, and Abraham A. Palmer. "The Genetics of Human Personality." *Genes, Brain, & Behavior* 17, no. 3 (2018). https://doi.org/10.1111/gbb.12439.

Santana, Eduardo J. "The Brain of the Psychopath: A Systematic Review of Structural Neuroimaging Studies." *Psychology & Neuroscience* 9, no. 4 (2016): 420-443.

Sapolsky, Robert M. *Behave: The Biology of Humans at Our Best and Worst.* Penguin Books (2017).

Shapiro, F. *Eye Movement Desensitization and Reprocessing (EMDR) Therapy: Basic Principles,*

Protocols, & Procedures (3rd ed.). New York, NY: The Guilford Press (2018).

Shulze, Lars, Isabel Dziobek, Alexander Vater, Hauke R. Heekeren, Mazen Bajbouj, Birgit Renneberg, Isabella Heuser, and Susanne Roepke. "Gray Matter Abnormalities in Patients with Narcissistic Personality Disorder." *Journal of Psychiatric Research* 47, no. 10 (2013): 1363-1369.

Simon Jr., George. *In Sheep's Clothing: Understanding and Dealing with Manipulative People*, rev. ed. Parkhurst Brothers Publishers (2010).

Simon Jr., George K. *Character Disturbance: The Phenomenon of Our Age.* Parkhurst Brothers Publishers (2011).

Sleep, Christopher E., Donald R. Lynam, and Joshua D. Miller. "Understanding Individuals' Desire for Change, Perceptions of Impairment, Benefits, and Barriers of Change for Pathological Personality Traits." *Personality Disorders: Theory, Research, and Treatment* 13, no. 3 (2021): 245-253.

South, Susan C., and Nicholas J. DeYoung. "The Remaining Road to Classifying Personality Pathology in the *DSM-5*: What Behavior

Genetics Can Add." *Personality Disorders: Theory, Research, and Treatment* 4, no. 3 (2013): 291-292.

Svrakic, Dragan M. "The Functional Dynamics of the Narcissistic Personality." *American Journal of Psychotherapy* 44, no. 2 (1990): 189-203.

Tavris, Carol. "The Scientist-Practitioner Gap: Revisiting 'A View from the Bridge' a Decade Later." In *The Science and Pseudoscience of Clinical Psychology*, edited by Scott O. Lilienfeld, Steven J. Lynn, and Jeffrey M. Lohr, 2nd ed., ix-xx. The Guilford Press (2015).

Torgersen, Svenn, Sissel Lygren, Per A. Oien, Siw I. Skre, Sigmund Onstad, Jorunn Edvardsen, Kristian Tambs, and Einar Kringlen. "A Twin Study of Personality Disorders." *Comprehensive Psychiatry* 41, no. 6 (2000): 416-425.

Torgersen, Svenn, John Myers, Ted Reichborn-Kjennerud, Espen Roysamb, Todd S. Kubarych, and Kenneth S. Kendler. "The Heritability of Cluster B Personality Disorders Assessed by Personal Interview and Questionnaire." *Journal of Personality Disorders* 26, no. 6 (2013): 848-866.

Trull, Timothy J., and Carey A. Durrett. "Categorical and Dimensional Models of Personality Disorders." *Annual Review of Clinical Psychology* 1 (2005): 355-380.

Weiss, Brendan, and Joshua D. Miller. "Distinguishing between Grandiose Narcissism, Vulnerable Narcissism, and Narcissistic Personality Disorder." In *Handbook of Trait Narcissism: Key Advances, Research Methods, and Controversies*, edited by Anthony D. Hermann, Amy B. Brunell, and Joshua D. Foster, 3-13. Springer International Publishing (2018).

Wilson, Nancy, Elaine Robb, Ruchit Gajwani, and Helen Minnis. "Nature and Nurture? A Review of the Literature on Childhood Maltreatment and Genetic Factors in the Pathogenesis of Borderline Personality Disorder." *Journal of Psychiatric Research* 137 (2021): 131-146.

Yakeley, Jessica. "Current Understanding of Narcissism and Narcissistic Personality Disorder." *BJPsych Advances* 24 (2018): 305-315.

Yudofsky, Stuart C. *Fatal Flaws: Navigating Destructive Relationships with People with Disorders of*

Personality and Character. American Psychiatric Association Publishing (2005).

Zimmerman, Mark. "Overview of Personality Disorders." September 2023. Accessed March 31, 2024. https://www.merckmanuals.com/professional/psychiatric-disorders/personality-disorders/overview-of-personality-disorders.

Other Titles by Peter Salerno

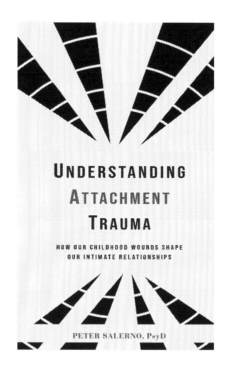

UNDERSTANDING ATTACHMENT TRAUMA

HOW OUR CHILDHOOD WOUNDS SHAPE OUR INTIMATE RELATIONSHIPS

PETER SALERNO, PsyD

Bonding with other people is necessary for survival.

But what happens when attachment fails? Nothing less than trauma. And this is much more common than you might think. Attachment trauma not only affects our daily lives, it also significantly impacts our most intimate relationships in the future.

Because understanding, healing and thriving go hand in hand, this book will not only help you understand how unresolved attachment trauma predicted your past relationship experiences, it is also a guide to healing wounds resulting from attachment trauma, to making sense of present relational difficulties, and in learning how to seek out safe and meaningful relationships in the future.

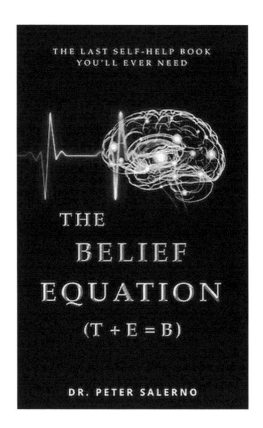

If you want to rid yourself of a negative mind-set that keeps you from getting what you want in life, THE BELIEF EQUATION will help you do just that.

Once we consciously program our mind and body to work in our favor rather than against our best interests, we no longer have to accept that we are somehow destined to be someone we don't want to be.

By learning to program yourself to influence your waking consciousness, as well as how you act and behave, you will be able to automatically live life on your terms without even having to think!

If there is any aspect of your life that you are currently not satisfied with, this book will show you how to get everything you want. All you have to do is solve The Belief Equation.

You become what you think about all day long.
-Ralph Waldo Emerson

Thoughts are so much more important than we give them credit for. Thoughts are powerful, and they have a strong influence over our emotions and our actions. Essentially, our thoughts can make us or break us.

Thought Detox condenses a wide range of simple yet brilliant truths and blends them into a concise daily guide accompanied by practical reflections that will help you to challenge thoughts and ideas that are not working for you and that might even be seriously impacting your mental wellbeing and physical health.

With this pocket guide of incredible thoughts and ideas at hand, you are well on your way toward developing better thinking habits that will change your perspective on life for the better while improving your overall health and wellbeing.

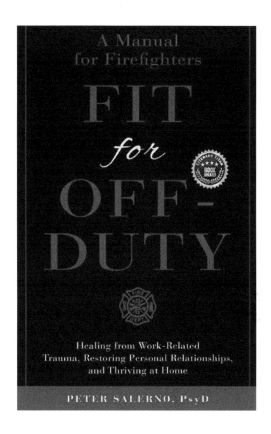

Every firefighter is a trauma survivor. Even veteran firefighters may not realize this. But constant exposure to traumatic events takes a serious toll. The body is affected, so is the nervous system, and so are the firefighter's personal relationships. Off-duty days can become something to dread rather than look forward to. It doesn't need to be this way. This book— written by a trauma therapist from a firefighter family—is a definitive manual for healing from trauma exposure for those who serve in the fire service and for those who love them. Feeling fit, healthy and unburdened by the effects of trauma can be a short-term therapeutic process. And there are steps firefighters can take on their own, immediately. **This book is a good place to start.**

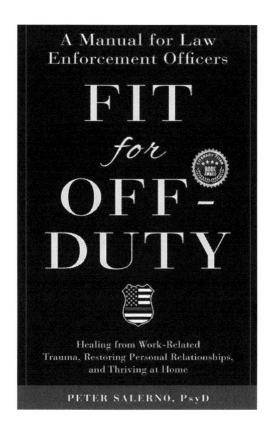

A Manual for Law
Enforcement Officers

FIT
for
OFF-
DUTY

Healing from Work-Related
Trauma, Restoring Personal Relationships,
and Thriving at Home

PETER SALERNO, PsyD

Every Law Enforcement Officer is a Trauma Survivor. Even veteran officers may not realize this. But constant exposure to traumatic events takes a serious toll. The body is affected, so is the nervous system, and so are the officer's personal relationships. Off-duty days can become something to dread rather than look forward to. It doesn't need to be this way. This book— written by a trauma therapist from a first responder family—is a definitive manual for healing from trauma exposure for those who serve in law enforcement and for those who love them. Feeling fit, healthy and unburdened by the effects of trauma can be a short-term therapeutic process. And there are steps law enforcement officers can take on their own, immediately. **This book is a good place to start.**

A Sneak Peek Inside
The Traumatic Cognitive Dissonance Workbook
by Peter Salerno, PsyD, LMFT

Available October of 2024

Introduction

When I became a therapist, I didn't intentionally set out to dedicate my entire career to deep diving into personality science and personality disorder research. I didn't set out to specialize in helping victims and survivors of **pathological relationship abuse** recover from the emotional devastation and hurt that inevitably result from a traumatic bond with a **Cluster B** personality. If you're reading this workbook, I think it's safe to assume that you have endured – or may currently be enduring – this very kind of devastation and hurt. The kind of devastation and hurt that hijacks your entire life and robs you of everything you thought you knew about yourself: your self-respect, your self-esteem, your self-worth, your self-confidence. And that's just scratching the surface.

My interest in the daunting task of writing on such a serious and painful subject sadly came from my own real life horror story, a story of survival and recovery from a relationship with a severely personality disordered partner, which forced me to come to terms with the reality that I had to let go of much of what I was formally taught and what I had always generally believed about human nature and behavior. Essentially, I had to accept the disturbing fact that there are a significant number of human beings who do really horrible things to really good people on purpose. These disordered individuals do not do these horrible things to good people because they were mistreated or because something bad happened to them; rather, these people do horrible things to others consciously and deliberately, because they want to.

Disordered personalities blend in with the crowd. They look, walk, and talk like everybody else, but their relational intentions are not like everyone else's. They exploit, manipulate, and violate other people because they believe they are entitled to do so. According to clinical research, almost twenty percent of the human population meet the diagnostic criteria for a personality disorder (Godwin & Lester, 2021). In other words, rather than standing out in the crowd, they make up a vast majority of the crowd. This is why it's so difficult to identify people with good intentions and people with not so good intentions. Relational "parasites" don't resemble the make-believe monsters underneath our beds or the distinctly ghastly characters we've grown accustomed to imagining from popular horror films. Relational parasites and predators resemble everyone else.

This workbook is designed to help you recover from the type of manipulation, coercion, and abuse that is consciously and deliberately inflicted by pathological individuals that are clinically referred to as **Cluster B** Personalities. These personality disorders are referred to as "cluster" disorders because the majority of individuals who meet the criteria for one cluster B disorder also meet the criteria for another cluster B disorder or come very close to it (Herpertz & Bertsch, 2022). The features of these four "dramatic and erratic" disorders tend to overlap quite frequently, so it's often exceedingly difficult for victims and survivors to truly know what kind of person they are dealing with – even if you've been married to a pathological personality for decades. And especially when you are caught in a web of chronic emotional abuse and truth manipulation.

If you are or have been faced with the unfortunate circumstance of being trapped in a relationship with a pathological personality, like a narcissist, an antisocial/psychopath, a borderline personality, or any other type of relational manipulator, this workbook intends to help you rid your life of these kinds of people and free you from the abuse once and for all. Unfortunately, the kind of trauma that comes from

pathological relationship abuse does not heal on its own or with passing time. That's also what this workbook is for.

When truth manipulation including pathological lying, gaslighting, subterfuge, and other forms of deceit and coercion are experienced on a chronic basis, the result is a devastating, disorienting, and bewildering form of trauma with a very unique indicator that goes far beyond typical trauma and stress-related symptoms. I call this unique indicator **Traumatic Cognitive Dissonance (TCD)**.

This Workbook and You

This workbook is divided into three sections and was created for victims of pathological relationship abuse so that you have a well-informed and accurate understanding of the psychological injuries you've suffered and endured, but also survived. It's important to remember that if you are reading this right now, you are in fact a survivor, and no one can take that away from you.

The first section explores how to recognize the signs that indicate you may be suffering from traumatic cognitive dissonance and did not even know it, or you did know it, but didn't know what to call it and you did not have any idea what to do about it. And maybe, neither does your current therapist or the previous professionals you sought answers from.

The second section will assist you in identifying how traumatic cognitive dissonance presents itself in your life, what happens to your brain and nervous system when subjected to chronic manipulation and abuse on the part of a cluster B personality, and how it has truly impacted you.

The third section will introduce you to exercises that are designed to help you rediscover and reunite with your true sense of self, the person you know yourself to be, the person buried underneath all the truth manipulation, coercion, subterfuge, and the other forms of abuse you were subjected to, but who is certainly not gone for good and has NOT been permanently damaged no matter what has happened or what you believe about yourself at this moment.

Available October of 2024

Made in United States
Troutdale, OR
09/15/2024

22830004R00147